D1603256

FUNDAMENTALS OF
Electric Waves

HUGH HILDRETH SKILLING, Ph.D.
Professor of Electrical Engineering, Stanford University

ROBERT E. KRIEGER PUBLISHING COMPANY
HUNTINGTON, NEW YORK
1974

Original Edition 1942, Second Edition 1948
Reprint 1974

Printed and Published by
ROBERT E. KRIEGER PUBLISHING CO., INC.
P.O. BOX 542
HUNTINGTON, NEW YORK 11743

Copyright © 1942, 1948 by
HUGH HILDRETH SKILLING
Reprinted by Arrangement

Library of Congress Catalog Card Number 74-8930
ISBN Number 0-88275-180-8

All rights reserved. No reproduction in any form of this book, in whole or in part (except for brief quotation in critical articles or reviews), may be made without written authorization from the publisher.

Printed in the United States of America

Library of Congress Cataloging in Publication Data

Skilling, Hugh Hildreth, 1905-
    Fundamentals of electric waves.

    1. Electric waves.   I. Title.
QC661.S63 1974   537.5   74-8930
ISBN 0-88275-180-8

PREFACE TO REPRINT EDITION

"It's easy to understand," say some.
"It gives a quick review," say others.
These are the kinds of comments that keep coming back to me, and I must say I value them.

The comments about the book being easy to understand come from students who are now working on electromagnetic theory. There is a difference between the very concept of field theory and that of circuits or networks or (in mechanics) of particles. What is gradient? What is curl? What is vector potential? What do Maxwell's equations really mean? The paddle-wheel concept of curl, for instance, turned out to be helpful to a good many readers who, like myself, wanted something easier to visualize than a set of partial differential equations. I take it as a high compliment when physics professors say (and they sometimes do), "... but if you want really to understand what it is all about, read Skilling's *Fundamentals*."

The comment about review comes typically from men who were once doctoral students and needed to prepare for an examination, or who wanted to look up a quantity that had escaped them. One excellent thing about fundamentals is that they do not grow old.

There are now ever so many more erudite books on electromagnetic theory, but it is not always evident that longer books are necessarily more lucid.

This little book on *Fundamentals* is arranged for those who may not have any previous knowledge of electromagnetic theory: all that is required is general college physics and mathematics through calculus. Vector analysis is introduced for use in the book, but it is not expected that the reader should have any earlier acquaintance with it.

The opening chapters of the book are concerned with electrostatics, the use of vector analysis, and similar matters that will be discouraging to the impatient reader. To some these opening chapters will be fascinating, to others, tedious. In any case they cannot be helped, for one cannot have waves without electric and magnetic fields, nor can one understand a wave until the basic fields are thoroughly familiar.

Maxwell's equations, appearing about the middle of the book, are presented as logical conclusions of the work that has gone before. Then, with their aid, radiation and wave propagation are readily developed, and these topics lead to a short discussion of antennas, transmission lines, and wave guides.

Problems are given with each chapter, and they are an integral part of the book. Most of them supplement some idea that is left without complete discussion in the text. They are arranged in the same order as the text material with which they are to be used, and in general one or two of them should be worked day by day. Also — and this is very important — many of the concepts of the book are new to the reader and will cease to seem strange only after continued and repeated use. Abraham and Becker, at the beginning of the examples in their *Classical Electricity and Magnetism*, refer the student to James 1 : 22, "But be ye doers of the word, and not hearers only, deceiving your own selves." I cannot think of any better advice.

Preparation of this book has left me indebted to many people. First is Hazel Dillon Skilling, my wife, whose name should properly appear on the title page as coauthor, except that she will not have it so. Stanford University has a farsighted and generous policy that encourages publication. This book is a result of that policy.

HUGH HILDRETH SKILLING

Stanford University
1974

Preface

This second edition extends the scope of *Fundamentals of Electric Waves*, giving more discussion of some of the most important applications and carrying theory somewhat further.

New material is included on wave guides. The chapter on antennas is rewritten. Discussion of reflection is substantially increased and reorganized. Wave propagation in semi-conducting media is considered. Propagation in ionized regions is discussed in a short chapter on the ionosphere.

The rationalized system of meter-kilogram-second units, the Giorgi system, is used throughout this edition. For the benefit of those who are more familiar with the centimeter-gram-second system, explanatory notes in the early chapters provide a transition. It is now clear that students will find mks units the most commonly used in current literature, and it appears that the sooner that system is learned the easier the students' work will be.

When the first edition was written, largely in 1941, a course in electromagnetic theory was rather a luxury for electrical engineers. Wartime developments quickly changed that view. There is no longer any need to explain in this preface the practical value of centimeter waves, or to tell of the importance of wave guides.

This book has been found effective for electrical engineering students at about the senior college year. It is arranged for those who do not necessarily have any previous knowledge of electromagnetic theory: all that is required is general college physics and mathematics through calculus. Vector analysis is introduced for use in the book, but it is not expected that the reader should have any earlier acquaintance with it.

The opening chapters of the book are concerned with electrostatics, the use of vector analysis, and similar matters that will be discouraging to the impatient reader. To some these opening chapters will be fascinating, to others, tedious. In any case they cannot be helped, for one cannot have waves without electric and magnetic fields, nor can one understand a wave until the basic fields are thoroughly familiar.

Maxwell's equations, appearing about the middle of the book, are presented as logical conclusions of the work that has gone before. Then, with their aid, radiation and wave propagation are readily developed, and these topics lead to a short discussion of antennas, transmission lines, and wave guides.

Problems are given with each chapter, and they are an integral part of the book. Most of them supplement some idea that is left without complete discussion in the test. They are arranged in the same order as the text material with which they are to be used, and in general one or two of them should be worked day by day. Also—and this is very important—many of the concepts of the book are new to the reader and will cease to seem strange only after continued and repeated use. Abraham and Becker, at the beginning of the examples in their *Classical Electricity and Magnetism*, refer the student to James 1 : 22; I cannot think of any better advice.

Preparation of this book has left me indebted to many people. First is Hazel Dillon Skilling, my wife, whose name should properly appear on the title page as coauthor, except that she will not have it so.

Stanford University has a farsighted and generous policy that encourages publication. This book is a result of that policy, administered by Chancellor Wilbur, President Tresidder, and Dean Terman.

For technical aid, also, I am indebted to Terman, and to many other authors including Stratton, Schelkunoff, Everitt, Fink, Albert, Harnwell, Guillemin, Ramo and Whinnery, Pierce, Page, Mason and Weaver, Abraham and Becker, King, Slater, Barrow, Ballantine, Carson, Chu, Southworth—too many to list.

Further reading in theory and in practice beyond the scope of this book is strongly recommended. The literature of electromagnetic theory as related to engineering is now so extensive that a bibliography is not even attempted. King gives a good list of books in *Electromagnetic Engineering* (McGraw-Hill Book Co., 1945, Vol. I, page 539). In addition to the standard texts and reference books, current periodical literature is often of particular interest and offers up-to-date lists of references.

I take the opportunity to express sincere thanks to professors who have advised me regarding this revision. I am grateful for the benefit of their experience with *Fundamentals of Electric Waves* in more than fifty schools. Their suggestions have guided me in preparing the new edition.

Hugh Hildreth Skilling

Stanford University
May 1948

Contents

CHAPTER

I.	Experiments on the Electrostatic Field	1
II.	Vector Analysis	10
III.	Certain Theorems Relating to Fields	38
IV.	The Electrostatic Field	51
V.	Electric Current	69
VI.	The Magnetic Field	77
VII.	Examples and Interpretation	99
VIII.	Maxwell's Hypothesis	109
IX.	Plane Waves	119
X.	Reflection	138
XI.	Radiation	160
XII.	Antennas	172
XIII.	Wave Guides	193
XIV.	Waves in the Ionosphere	228
	Index	241

TABLE

I.	Units and Symbols	*Inside Front Cover*
II.	Formulas and Theorems of Vector Analysis	*Facing Inside Back Cover*
III.	Electromagnetic Equations and Wave Relations	*Inside Back Cover*
IV.	Electrodynamic Potentials	161
V.	Earth Characteristics	184
VI.	Components of Transverse Electric Waves in Rectangular Guides	211
VII.	Components of Transverse Magnetic Waves in Rectangular Guides	212
VIII.	Auxiliary Formulas for Waves in Rectangular Guides	212
IX.	Certain Modes in Cylindrical Guides	217

CHAPTER I

Experiments on the Electrostatic Field

Fields. The study of electricity commonly begins with electric circuit theory. Current is considered to flow in a wire, being driven through resistance, inductance, and capacitance by the appropriate voltages. This is the natural approach to the subject, for electric circuits are tangible, and to most people they are reasonably familiar.[1] The historical development of the subject, however, was quite the opposite: magnetic and electrostatic *fields* were well understood before circuit theory was developed—before even Ohm's law was discovered. Logically, also, as will later be seen, circuit theory may be considered as a special case of the more general theory of electromagnetic fields.

There are various kinds of fields. There are vector fields and scalar fields. A gravitational field, for example, is a vector field. Consider the gravitational field within a room. If an object of unit mass were placed at any point of space within the room, there would be a force upon it. This is a particularly simple example of a field of force, for in it the force is practically the same at every point within the room; it is the same in magnitude and vertically downward in direction. It is a *vector* field, for it is not fully defined until it is known at all points in both magnitude and direction.

Fields of force are always vector fields. A *scalar* field may be illustrated by temperature. A temperature field would be determined if one were to measure the temperature at each point in a room with a thermometer. There is a value of temperature at each point, but no direction is associated with temperature. The temperature field is therefore a *scalar* field.

Electric and magnetic fields are vector fields. The electrostatic field will be considered first, for it is in some ways the simplest.

EXPERIMENT I. *The Electric Field.* It is found by experiment that there is a field of force about any object that has an electric charge. This field of force is made evident when an exploring particle that carries

[1] This approach is used in *Transient Electric Currents*, H. H. Skilling, McGraw-Hill Book Co. New York, 1937.

1

on itself a small electric charge is placed at some point in the region near the charged body. If it is placed at point a in Fig. 1, there is a force $\mathbf{F}_a$; at points b and c, there are forces $\mathbf{F}_b$ and $\mathbf{F}_c$. If the charge on the exploring particle is changed, force upon the particle changes in proportion. This experiment, which will be called Experiment I, makes it possible to give the following definition of the electrostatic field.

By definition, the **electrostatic field strength** at each point is equal in magnitude and direction to the force exerted on a small exploring particle carrying unit charge of positive electricity that is at rest at that point.

Symbolically,

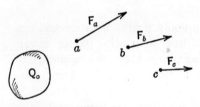

Fig. 1

$$\mathbf{F} = \alpha Q \mathbf{E} \qquad [1]$$

where $\mathbf{F}$ and $\mathbf{E}$ are force and electric field strength, respectively, and Q is the charge on the exploring particle. Note that this equation is not limited to any one point but applies at *all* points. It is therefore a *field* equation, and, since it relates both magnitude and direction, it is a *vector* field equation.

The quantity α is merely a factor of proportionality, and its value depends on the units used. Proper definition of the unit of electric charge and the unit of electric field strength will make $\alpha = 1$, so that

$$\mathbf{F} = Q\mathbf{E} \qquad [2]$$

Thus, Q may be measured in statcoulombs and $\mathbf{E}$ in statvolts per centimeter, and $\mathbf{F}$ will be in dynes in equation 2; or Q may be in coulombs, $\mathbf{E}$ in volts per meter, and $\mathbf{F}$ will be in newtons.

Equation 2 defines the electric field.

Having defined the electric field, we may study its properties. How does it arrange itself in space? For this purpose, more experimentation is necessary. Three more experiments with the exploring particle will be considered.

These experiments are not suggested as practical experiments to be done in the laboratory. They would be difficult to perform. But they are exceptionally useful experiments to serve as a foundation for theory. Let us accept, for purposes of this discussion, that the following experiments have been performed with the results given below.

EXPERIMENT II. An exploring particle, which carries a small electric charge, is moved through a region in which there is an electrostatic field. It is found that, when the particle is moved in a closed path so that it returns to the point from which it started, no total work is done either on the particle or by the particle.

In Fig. 2, for example, an exploring particle a may be moved around the path indicated by the dash line. While the particle recedes from the charged body from which the electric field radiates (as indicated by arrows **E**), work is done by the electric field upon the particle. But, as the particle returns toward the charged body, following the other half of the path indicated, it must do equal work in moving against the force of the field.

It will be noted that this conclusion is in agreement with the principle of conservation of energy. If the particle returned to its initial point with an excess of energy, it could go around again and continue to go around, each time gaining a little energy without a corresponding loss of energy in another part of the system. This would make perpetual motion feasible and is contrary to the principle of conservation of energy. It is equally impossible that the particle should return to its initial point with a deficiency of energy, for (assuming no friction) the total energy of the system would then have diminished.

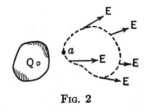

Fig. 2

The conclusion from Experiment II is entirely independent of the shape of the path followed by the exploring particle; it may be circular, elliptical, square, or any other closed path. The conclusion is also independent of the source of the electric field, which may emanate from a charged body, or from a number of charged bodies, or from a charge that is diffuse in space. The charge that produces the electric field must not change in any way while the exploring particle is making its complete circuit, for this is an electro*static* experiment; and it follows that the charge on the exploring particle must be so small that its presence does not appreciably alter the distribution of the main charge.

Since energy is equal to the product of force and distance, and the net energy is the summation, or integral, of the individual energies contributed by each increment of distance around the closed path, it follows that

$$\oint \mathbf{F} \cdot d\mathbf{s} = 0 \qquad [3]$$

Then, since the force field and the electric field are related by a constant factor, as in equation 2, we obtain

$$\oint \mathbf{E} \cdot d\mathbf{s} = 0 \qquad [4]$$

This expression is a line integral, **s** representing distance along the path of integration. The small circle superimposed upon the integral sign indicates that the integration is to be carried out around a *closed* path. The notation will be further explained in Chapter II.

EXPERIMENT III. A closed surface is located in space. It may be any shape: spherical, ellipsoidal, cubical, or irregular, but it must be completely closed and must not be, for example, a sphere with a hole in it. It is a purely imaginary surface and is used only to isolate the space within from the space outside.

Let us choose the imaginary closed surface so that it does not pass through any solid or liquid material. (This is a restriction that will be removed later, but at present it simplifies the discussion.) It would be better from the theoretical point of view if the imaginary surface did not pass through any material substance, including air; but air affects the results of Experiment III by less than a tenth of a per cent and is usually negligible.

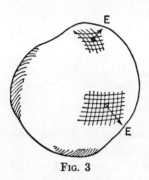

FIG. 3

Experiment III is now performed by measuring the electric field strength (by means of an exploring particle) at every point on the closed surface. This can best be done by dividing the surface into a very large number of small sections, as in Fig. 3, each having an area *da*. The component of the electric field normal to the small area *da* is then determined. If the normal component is outward it is called positive, if inward it is negative. Then all the normal components are multiplied by their respective areas, and the results are added together (as in the left-hand member of equation 5). When this experiment has been tried for all possible surfaces under the widest imaginable variety of circumstances, the conclusion is reached that the summation described above is proportional to the amount of electric charge enclosed within the surface on which measurements are made. If ϵ_0 is a constant and Q is the amount of charge within the surface,

$$\epsilon_0 \oint \mathbf{E} \cdot d\mathbf{a} = Q \quad \text{(in vacuum)} \qquad [5]$$

This is a surface integral, **a** representing area. The small circle upon the integral sign here indicates that integration is to be carried out over a *closed* surface.

If, in Fig. 3, the summation of the normal component of **E** over the entire surface is zero, it follows that there is no electric charge within the surface or, if there is any positive charge within, there is also an equal amount of negative charge. If the summation is not zero, there is a net electric charge within the surface, and the amount of contained charge is proportional to the summation of the electric field strength over the surface, as in equation 5.

The value of ϵ_0 depends on the units in which electric field strength, area, and charge are measured.[2] Using "practical" units, with **E** in volts per meter, **a** in square meters, and Q in coulombs, ϵ_0 is very nearly $\dfrac{10^{-9}}{(9 \times 4\pi)}$, or a little more nearly 8.855×10^{-12}.

EXPERIMENT IV. Now let us repeat Experiment III, this time taking measurements of electrostatic field strength at points on an imaginary surface in oil. As before, the field strength is measured by determining the force on a charged exploring particle, but, whereas in Experiment III the charged particle was in air (or, strictly, in vacuum), now each measurement is to be made with the particle in oil. The value of the integral of equation 5, corresponding to a certain amount of electric charge within the enclosing surface, proves to be different from the value obtained when the same amount of electric charge was contained within a closed surface in air.

In petroleum oil, the experimental value obtained for the integral of equation 5 is about half the value in air. If the experiment is repeated in other substances, other different values will be found. To make equation 5 apply generally for all substances, it is necessary to include in the equation a factor that is characteristic of the material in which measurements are made. This will be written κ and is called the **relative dielectric constant** (sometimes the "specific inductive capacity") of the material.

---

[2] The common unit systems for electrostatic work are the "electrostatic" cgs (centimeter-gram-second) system using the statvolt, statampere, statohm, etc., and the "practical" mks (meter-kilogram-second) system employing volts, amperes, and ohms. All equations of this chapter apply equally well with either system. Units are discussed further on page 8.

Another choice affecting units that must be made at this time is between a "rationalized" and an "unrationalized" system of equations. These differ by a factor of 4π in defining ϵ_0. Equation 5 is written "rationalized"; the "unrationalized" form is

$$\epsilon_0 \oint \mathbf{E} \cdot d\mathbf{a} = 4\pi Q \qquad [5a]$$

"Rationalized" equations may use either cgs or mks units; so may "unrationalized" equations. The "unrationalized" cgs system is the Gaussian. The "rationalized" mks system is the Giorgi system. The numerical value of ϵ_0 in equation 5a in the Gaussian system is unity.

Then
$$\kappa\epsilon_0 \oint \mathbf{E} \cdot d\mathbf{a} = Q \qquad [6]$$

The factor κ is called *relative* because it shows how much less the electrostatic forces are in a given material than in empty space. The value of κ depends mainly on the nature of the material, but it changes somewhat with temperature and other physical conditions. For most oils that are derivatives of petroleum its value is between 2 and 2.5. In cottonseed or olive oil it is about 3. In ethyl alcohol at room temperature it is about 25, and in pure distilled water about 80. In empty space κ is 1, of course, and in air about 1.0006.

To conclude Experiment IV, we make measurements on surfaces that pass through various different substances. The surface of integration may thus be partly in air and partly in oil. The value of κ must then be changed as we pass from one material to another, and equation 6 is more properly written

$$\oint \kappa\epsilon_0 \mathbf{E} \cdot d\mathbf{a} = Q \qquad [7]$$

for κ varies during the process of integration.

Frequently the constant κ, which has to do with the characteristics of the material in the electric field, and the constant ϵ_0, which takes care of dimensions and units, are combined into a single constant, ϵ. Then,[3] with $\epsilon = \kappa\epsilon_0$,

$$\oint \epsilon \mathbf{E} \cdot d\mathbf{a} = Q \qquad [8]$$

Electrostatic Flux. By way of introducing electrostatic flux, consider a vector $\mathbf{D}$ defined as

$$\mathbf{D} = \epsilon \mathbf{E} \qquad [9]$$

When this new symbol is used, equation 8 becomes

$$\oint \mathbf{D} \cdot d\mathbf{a} = Q \qquad [10]$$

We are interested in the product of $\mathbf{D}$ and area. $\mathbf{D}$ is called **electric flux density,** and the product of $\mathbf{D}$ and area is called **electric flux.**

By definition

$$\text{Electrostatic flux} = \int \mathbf{D} \cdot d\mathbf{a} \qquad [11]$$

---

[3] In Gaussian units, with $\epsilon_0 = 1$ and the unrationalized equation 5a;
$$\oint \kappa \mathbf{E} \cdot d\mathbf{a} = 4\pi Q$$

(The notation of this expression is explained in the next chapter, but its meaning is clear from this discussion.) In Fig. 4 the flux-density field is shown normal to a surface, and the flux passing through the surface is simply D times area. If the field is not normal to the surface, flux is equal to the product of area and the *normal component* of the field. In the extreme case, with **D** parallel to the surface, there is no normal component and no flux penetrates the surface.

With this definition of flux, the left-hand member of equation 10 is flux passing through the closed surface of integration. The equation,

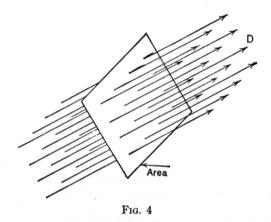

Fig. 4

then, is a mathematical formulation of the following statement: the flux passing through a closed surface is equal [4] to the electric charge contained within the surface.

Lines of flux can be drawn (or at least imagined) emanating from positive charge, passing through space in the direction of the electric field, and terminating on negative charge. If each line represents a unit amount of flux, there will be one line issuing from each unit of positive charge, and one line terminating on each unit of negative charge. In space where there is no electric charge, the flux lines must be continuous, for they cannot terminate.

When lines of flux pass from a charge $+Q$ to a charge $-Q$, as in Fig. 5, all the flux lines will penetrate any surface that completely surrounds the charge $+Q$. Such a closed surface is indicated by the dash line a. By counting the flux lines that pass through this surface, it is possible to know how much charge is within the surface. Lines going outward

---

[4] In Gaussian units $\oint \mathbf{D} \cdot d\mathbf{a} = 4\pi Q$, and 4π flux lines emanate from one unit of charge.

are counted as positive, lines inward as negative. Hence, within the surface b there is no charge, for the algebraic sum of lines through the surface is zero. Within c the charge is negative. Within d there is no net charge; the algebraic sum of lines penetrating the surface is zero, and the amount of positive charge within the surface is equal to the amount of negative charge.

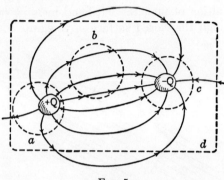

Fig. 5

Units. Besides Q, the charge, five electrical quantities have now been introduced. It will be well to review them.

- **E** is electric field strength, a vector quantity, determined by the force on a charged exploring particle. In practical units it is measured in volts per meter.
- **D** is electric flux density, a vector quantity, determined by the direction and density of flux lines that emanate from electric charge and follow the direction of the electric field. In practical units it is measured in coulombs per square meter.
- ϵ is the ratio of **D** to **E**; it is different for different materials. It is called absolute dielectric constant or permittivity.
- ϵ_0 is the ratio of **D** to **E** in vacuum; in practical units it is 8.855×10^{-12}. It is called the absolute dielectric constant, or permittivity, of free space.
- κ is ϵ/ϵ_0. It is called the relative dielectric constant, or specific inductive capacity, and is characteristic of the dielectric material.

"Practical units" as referred to in this chapter means the system including the volt, ampere, coulomb, farad, henry, ohm, watt, and ampere-turn, with distance in meters, mass in kilograms, and time in seconds. The unit of work is the joule (or watt-second), and the unit of force is the newton (or joule per meter). One newton is 10^5 dynes, a convenient force of about 102.0 grams or a little over $3\frac{1}{2}$ ounces (this book weighs about 6 newtons). This system, comprising the practical electrical units, is the rationalized mks or Giorgi system.

The subject of units and dimensions is a fascinating one.[5] As every author is inclined to introduce a few novelties to suit his personal preference, the subject is extremely complex. However, most electrical engineering literature on electromagnetic theory now uses practical units in the rationalized mks or Giorgi system, with or without the author's individual modifications, and the mks system has been recommended by various international conferences.[6]

The outstanding advantage of the mks system is that it uses familiar volts, amperes, ohms, and other electrical units. The disadvantage is that peculiar values are assigned to the dielectric constant and permeability of free space. Inconveniences occur in all systems, but the Giorgi system is cleverly arranged to have the powers of 10, the factors of 3×10^{10}, and most of the factors of 4π all bound together in ϵ_0 and μ_0; once the odd values of these constants have been mastered, the petty annoyances are largely over.

Information about electrostatic fields that will be needed in later chapters can be deduced from equations 2, 4, and 8. This is done by mathematical methods, and the most convenient mathematics to use is vector analysis. It is desirable, therefore, to introduce some of the general mathematical relations of vector analysis before going on to further study of the electric field.

PROBLEMS

1. A body carrying a positive electric charge of 1000 micromicrocoulombs (1000×10^{-12}) is in an electric field of 5000 volts per centimeter. What is the electric force on the body in newtons? In milligrams?

2. Electric field strength is measured at all points of a spherical surface of 10-centimeter radius in air. It is found to be everywhere normal to the surface, 10,000 volts per meter in magnitude, and directed outward. How much electric charge (in microcoulombs) is contained within the spherical surface?

3. What quantity of electric flux comes out of the spherical surface of Problem 2? What quantity would come out of the same charge in petroleum oil? What would be the value of **E** at the surface in petroleum oil?

---

[5] For an excellent summary see "Physical Units and Standards" by Ernst Weber, Section 3 of *Handbook of Engineering Fundamentals*, John Wiley & Sons, New York, 1936.

[6] "I.E.C. Adopts MKS System of Units," Arthur E. Kennelly, *Trans. AIEE*, Volume 54, 1935, pages 1373–1384.

"Recent Developments in Electrical Units," Arthur E. Kennelly, *Electrical Engineering*, Volume 58, February, 1939, pages 78–80.

"Revision of Electrical Units," E. C. Crittenden, *Electrical Engineering*, Volume 59, April, 1940, pages 160–163.

CHAPTER II

Vector Analysis

Vector Multiplication. Vectors are useful for various purposes. Force can be represented by a vector. So can distance. If a force **F** acts on a body while that body is moving through a distance **s**, as in Fig. 6a, the work done by the force is the product of force and distance. But it is not the simple algebraic product, for the angle between the direction of the force and the direction of travel is important. If the magnitude of the force is represented by F, the magnitude of the distance by s, and the angle between their directions is θ, then the work done is

$$W = Fs \cos \theta \qquad [12]$$

Note that the vector quantities for force and distance are written **F** and **s**, whereas their scalar magnitudes are F and s; this sytem of notation is rather generally adopted and will be used consistently in the present discussion.

Since the type of multiplication indicated in equation 12 is quite common in physical problems, it is given a special symbol in vector analysis: when two vectors are written with a dot between them, it is an indication of multiplication of this type. Hence, equation 12 may be written

$$W = \mathbf{F} \cdot \mathbf{s} \qquad [13]$$

In the general case of any two vectors **A** and **B**, the so-called "scalar product" or "dot product" is defined as follows:

$$\mathbf{A} \cdot \mathbf{B} = AB \cos \theta \qquad [14]$$

As in the case of work, in equation 13, this kind of product is always a scalar quantity, although the quantities multiplied together are both vectors. It is for this reason that it is called the **scalar product.**

There is also another type of multiplication commonly encountered in physical problems. The simplest example is the computation of area, as in Fig. 6b, where two vectors **A** and **B** are shown as the sides of a parallelogram. The area of the parallelogram is

$$\text{Area} = AB \sin \theta \qquad [15]$$

The same type of multiplication is encountered in finding the force on a conductor carrying current in a magnetic field. In Fig. 6c, current of I amperes (a *scalar* quantity) is flowing in a conductor the direction and length of which are represented by a *vector* **L**. The magnetic field is represented in magnitude and direction by the vector **B**. Then the force on the conductor will be the vector quantity **F** which is perpen-

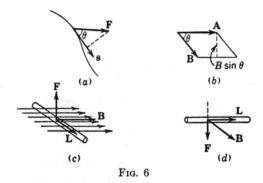

Fig. 6

dicular to both **L** and **B**. This defines its direction, and its magnitude is given by

$$F = ILB \sin \theta \qquad [16]$$

The sense of the force is upward in the figure, in accordance with the rule for force in a magnetic field.

Because this type of multiplication is quite common, it also is given a special symbol. The vectors **L** and **B** are written with a cross between them so that equation 16 is written

$$\mathbf{F} = I\mathbf{L} \times \mathbf{B} \qquad [17]$$

This type of operation gives what is known as the cross product or **vector product**. The latter name comes from the fact that the result of this type of multiplication, such as the force in equation 17, is itself a vector.

It is not at once apparent that the area of Fig. 6b is a vector quantity. But a surface obviously does have an orientation in space, and, by convention, an area is represented by a vector whose direction is *normal* to the surface, and with length proportional to the area. Hence the area of Fig. 6b is represented by a vector perpendicular to the plane of the paper.

A question naturally arises regarding the sense of the resultant vector, such as **F** in equation 17. What is there in the equation to signify whether the force is upward or downward? If, for example, **L** and **B** were to be interchanged, as indicated in Fig. 6d, the magnitude of the force would be unchanged, but the sense would be reversed and would become downward. To avoid ambiguity in the mathematical statement of such a problem, the vector product is so defined that the sense of the resultant vector is indicated by the order in which the two component vectors are written.

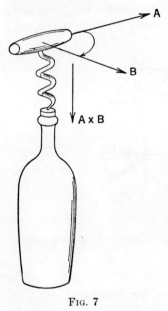

Fig. 7

This is a useful and thoroughly satisfactory means of defining the direction of the vector product. Yet expressing in words the defining relation is somewhat awkward. It is customary to remember the relation by a certain arrangement of fingers and thumb on the right hand, or in terms of the rotation of a so-called "right-hand" screw thread.

The product of two vectors, **A** × **B**, is itself a vector of magnitude $AB \sin \theta$, in direction normal to the plane that contains both **A** and **B**, and of such sense that, if a right-hand screw (see Fig. 7) were rotated from **A** to **B** (through the angle that is less than 180 degrees), it would screw in the direction of the product.

Following this rule, equation 17 is seen to give the proper direction of force in either Fig. 6c or 6d, and the vector **A** × **B** which represents area in Fig. 6b is properly *into* the sheet of paper. Note, however, that the product **B** × **A** is, by the same rule, a vector directed *outward* from the sheet of paper; this is merely an illustration of the general rule that

$$\mathbf{B} \times \mathbf{A} = -(\mathbf{A} \times \mathbf{B}) \qquad [18]$$

It will be seen that this operation does not follow the fundamental commutative law of ordinary algebra which says that $ab = ba$. It is natural to question what justification there can be for denying a fundamental law of algebra, and a short discussion may be helpful in this connection.

All the operations of algebra, including multiplication, are defined for use with numbers. The rule for multiplication is particularly easy for integers; for instance, 7 times 5 is five 7's added together (or, by the commutative law, it is also the sum of seven 5's). This rule is extended quite readily to the multiplication of fractions, and by an additional convention regarding sign it can be made to serve for negative numbers also. But it simply has no meaning if one tries to apply it to vectors. There are, however, certain operations so commonly performed upon vectors that it seems desirable to give them names; two of these operations are discussed above, and the confusing thing about the situation is that they are *both* called multiplication. Actually, it is very doubtful if either should properly be called multiplication; probably that name should be reserved for the algebraic product of two *scalar* quantities, and completely new names could then be assigned to the operations upon vectors that are known (however improperly) as vector multiplication leading to the scalar product, in one case, and to the vector product, in the other. But the nomenclature is so well established that it cannot be avoided.[1]

It now becomes clear that, since the vector product is not algebraic multiplication but is defined quite independently, it is not constrained to follow algebraic laws. It is not surprising that the commutative law fails to apply to the vector product. Rather, it is to be remarked that the commutative law does apply in the case of the scalar product, as defined in equation 14.

Unit Vectors. No single expression has yet been given to define the vector product. The following may be used, although it requires a short explanation:

$$\mathbf{A} \times \mathbf{B} = \mathbf{n} AB \sin \theta \qquad [19]$$

In this expression, A and B are the scalar magnitudes of $\mathbf{A}$ and $\mathbf{B}$, θ is the angle between $\mathbf{A}$ and $\mathbf{B}$, and $\mathbf{n}$ is a vector of unit length in a direction normal to both $\mathbf{A}$ and $\mathbf{B}$ and with sense, as defined above, forming a right-hand system with $\mathbf{A}$ and $\mathbf{B}$.

The right-hand side of equation 19 illustrates a method of describing a vector. The magnitude of the vector product is given by $AB \sin \theta$, but this expression does not give direction. The unit vector $\mathbf{n}$ serves

---

[1] Yet another kind of multiplication, so called, is defined for use with complex quantities. This is the operation according to which

$$(Ae^{i\alpha})(Be^{i\beta}) = ABe^{i(\alpha+\beta)}$$

This law is familiar to students of alternating-current phenomena, for it is very widely used in connection with a convention that makes it possible to represent real quantities that vary sinusoidally with time by means of complex quantities. The complex quantities, which are themselves scalar, can then be represented by rotating vectors in the complex plane. This may be considered a third type of multiplication of vectors, entirely different from either of the other two.

to define a direction normal to **A** and **B** and thereby specifies the direction of the vector product. This artifice of employing a unit vector [2] to give direction is very often useful.

The most common use of unit vectors to define direction is in connection with coordinate axes. In Fig. 8, three axes are shown, marked X, Y, and Z. The position of any point in space may be defined by reference to such axes, in the familiar manner of analytic geometry, using Cartesian coordinates. Similarly, the length and direction of any vector can be expressed by giving the projections of the vector upon the three axes. Any vector **A** may be described as being made up of three mutually perpendicular components (as in Fig. 8): A_x in the x direction, A_y in the y direction, A_z in the z direction.

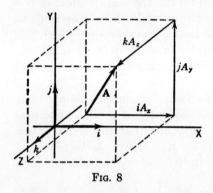

Fig. 8

Now A_x is a scalar quantity. It is the *length* of the x component of **A**. A_y and A_z are also scalar quantities. It is consequently not correct to say that the vector **A** is equal to the sum of A_x, A_y, and A_z. But it is correct to say that **A** is equal to the sum of a vector in the x direction of length A_x, a vector in the y direction of length A_y, and a vector in the z direction of length A_z.

Such a statement of equality is cumbersome, but it can be simplified by defining three unit vectors as follows: The vector **i** is a vector of unit length in the x direction (see Fig. 8); the vector **j** is a vector of unit length in the y direction; the vector **k** is a vector of unit length in the z direction. Now, when it becomes necessary to write of "a vector in the x direction of length A_x," it is only necessary to set down "$\mathbf{i}A_x$," which expresses exactly the same idea. So it is correct to write

$$\mathbf{A} = \mathbf{i}A_x + \mathbf{j}A_y + \mathbf{k}A_z \qquad [20]$$

and this notation will be used frequently.

Products involving these unit vectors are of frequent occurrence and deserve special consideration. Consider the dot product **i · i**; both **i**'s

---

[2] It should be noted that no equation can be correct unless either both sides are scalar quantities or both sides are vector quantities. A vector cannot be equated to a scalar. Hence equation 19 could not be correct in the absence of the symbol **n**, for without it the left-hand member would be a vector and the right-hand member a scalar.

UNIT VECTORS

are of unit length and the angle between them is zero; therefore, by equation 14, the product is unity, a scalar value. But consider $\mathbf{i} \cdot \mathbf{j}$; the angle between these two is 90 degrees; hence by equation 14 their product is zero. Physically, the projection of $\mathbf{i}$ upon $\mathbf{i}$ is unity, and the projection of $\mathbf{i}$ upon $\mathbf{j}$ is zero.

Consider the vector product $\mathbf{i} \times \mathbf{j}$; by equation 19 the product will be a vector of unit length normal to both $\mathbf{i}$ and $\mathbf{j}$, and in the direction that would be taken by a right-hand screw while being rotated from $\mathbf{i}$ to $\mathbf{j}$. A moment's study of Fig. 8 shows that this product is identical with $\mathbf{k}$. Hence $\mathbf{i} \times \mathbf{j} = \mathbf{k}$. But note that $\mathbf{j} \times \mathbf{i} = -\mathbf{k}$.

Because the angle between two similar unit vectors is zero, it is apparent from equation 19 that $\mathbf{i} \times \mathbf{i}$ is zero. This is illustrated by the fact that the area of a parallelogram, as in Fig. 6b, approaches zero as the two adjacent sides approach each other.

A partial tabulation of products of unit vectors follows:

$$\mathbf{i} \cdot \mathbf{i} = 1 \qquad \mathbf{i} \times \mathbf{i} = 0$$
$$\mathbf{i} \cdot \mathbf{j} = 0 \qquad \mathbf{i} \times \mathbf{j} = \mathbf{k} \qquad [21]$$
$$\mathbf{i} \cdot \mathbf{k} = 0 \qquad \mathbf{i} \times \mathbf{k} = -\mathbf{j}$$

Coordinate systems are essential in connection with vector analysis, and it will be helpful to express some of the more important vector operations in terms of Cartesian components. (A right-hand system of rectangular coordinates, as in Fig. 8, is used.)

The scalar product of any two vectors $\mathbf{A}$ and $\mathbf{B}$ may be expanded into

$$\mathbf{A} \cdot \mathbf{B} = A_x B_x + A_y B_y + A_z B_z \qquad [22]$$

To prove that this is true, substitute equation 20 for $\mathbf{A}$ and a similar expression for $\mathbf{B}$, and multiply term by term:

$$\mathbf{A} \cdot \mathbf{B} = (\mathbf{i} A_x + \mathbf{j} A_y + \mathbf{k} A_z) \cdot (\mathbf{i} B_x + \mathbf{j} B_y + \mathbf{k} B_z)$$
$$= \mathbf{i} \cdot \mathbf{i} A_x B_x + \mathbf{j} \cdot \mathbf{j} A_y B_y + \mathbf{k} \cdot \mathbf{k} A_z B_z$$
$$+ \mathbf{i} \cdot \mathbf{j} A_x B_y + \mathbf{i} \cdot \mathbf{k} A_x B_z + \mathbf{j} \cdot \mathbf{i} A_y B_x + \mathbf{j} \cdot \mathbf{k} A_y B_z$$
$$+ \mathbf{k} \cdot \mathbf{i} A_z B_x + \mathbf{k} \cdot \mathbf{j} A_z B_y \qquad [23]$$

The first three terms of this expansion give the right-hand member of equation 22, for the dot products of identical unit vectors are unity, and the other six terms disappear because all the dot products of the unlike unit vectors are zero.

As a physical illustration of the expansion of the scalar product in equation 22, consider the product of the force **F** and the distance **S** as shown in Fig. 9. The coordinate axes are selected in such a way that both force and distance lie in the X–Y plane, so the problem is merely a two-dimensional one. The vectors shown have the components F_x and F_y, and S_x and S_y. It is physically evident that the component of force F_x acting through the distance S_y does not represent any work, for the force and distance are perpendicular. The same is true for the force F_y acting through the distance S_x. But the force F_x applied

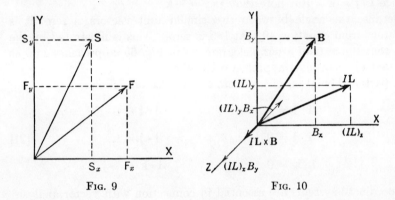

Fig. 9   Fig. 10

through the distance S_x results in work equal to $F_x S_x$, and the other components give $F_y S_y$. These latter terms, consequently, are retained in the scalar product, the full three-dimensional form of which is given in equation 22, while products of components along different axes (the zero terms of equation 23) contribute nothing to the scalar product.

The vector product, or cross product, of any two vectors may be expanded in a similar manner:

$$\mathbf{A} \times \mathbf{B} = \mathbf{i}(A_y B_z - A_z B_y) + \mathbf{j}(A_z B_x - A_x B_z) + \mathbf{k}(A_x B_y - A_y B_x) \quad [24]$$

The proof of this is exactly parallel to the proof of equation 22.

A physical illustration of the significance of terms in equation 24 is found in considering a current element (IL) in a magnetic field **B**, and the force $I\mathbf{L} \times \mathbf{B}$ that results upon the current-carrying conductor. Select coordinates so that both $(I\mathbf{L})$ and **B** lie in the X–Y plane, as in Fig. 10. The current component $(IL)_x$ and the magnetic field component B_x do not react, for there is no force on a current parallel to a magnetic field. But $(IL)_x$ and B_y, being at right angles, react to give the force $(IL)_x B_y$ in the direction of the Z axis. And $(IL)_y$ and B_x produce a force of magnitude $(IL)_y B_x$ that is directed along the Z axis in the negative direction. Total force consequently is the algebraic sum of these two, or

$$(IL)_x B_y - (IL)_y B_x \qquad [25]$$

This expression will be seen to correspond to the last term of equation 24. The other terms of equation 24 appear if (IL) and $\mathbf{B}$ have Z components also. Products of components with like subscripts, such as $(IL)_x B_x$, or $(IL)_y B_y$, do not appear in the expansion of the vector product.

It is convenient to express the expansion of the vector product as a determinant, and it makes it easier to remember.

$$\mathbf{A} \times \mathbf{B} = \begin{vmatrix} \mathbf{i} & \mathbf{j} & \mathbf{k} \\ A_x & A_y & A_z \\ B_x & B_y & B_z \end{vmatrix} \qquad [26]$$

If this determinant is expanded according to the ordinary rules (which may be found in algebra books or in the mathematical sections of handbooks), it is identical with equation 24.

Triple Products. The vector product of two vectors is itself a vector, and its product with some other vector may be found. The second multiplication may be either a scalar or a vector product, thus:

$$(\mathbf{A} \times \mathbf{B}) \cdot \mathbf{C} \qquad [27]$$

or

$$(\mathbf{A} \times \mathbf{B}) \times \mathbf{C} \qquad [28]$$

These are obviously different, and must be considered one at a time.

First consider expression 27. If $\mathbf{A}$, $\mathbf{B}$, and $\mathbf{C}$ are any three vectors, as in Fig. 11, the scalar triple product of expression 27 is the volume of the parallelopiped shown, of which $\mathbf{A}$, $\mathbf{B}$, and $\mathbf{C}$ are the three edges. This is evident when it is realized that $\mathbf{A} \times \mathbf{B}$ is the area of the top of the parallelopiped, and $(\mathbf{A} \times \mathbf{B}) \cdot \mathbf{C}$ is the product of this area and the normal component of the edge $\mathbf{C}$.

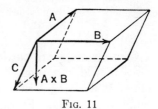

Fig. 11

The volume of the same parallelopiped will be found when the area of the side bounded by $\mathbf{B}$ and $\mathbf{C}$ is multiplied by the normal component of $\mathbf{A}$, and consequently

$$(\mathbf{A} \times \mathbf{B}) \cdot \mathbf{C} = (\mathbf{B} \times \mathbf{C}) \cdot \mathbf{A} \qquad [29]$$

and similarly either of these is equal to $(\mathbf{C} \times \mathbf{A}) \cdot \mathbf{B}$.

18 VECTOR ANALYSIS

This scalar triple product is sometimes written [ABC], and this is an adequate notation, for a little study will show that the vectors may be multiplied in any order provided the cyclic order ABC be retained, but that

$$[ABC] = -[CBA] \qquad [30]$$

The dot and cross may be inserted as desired in expression 30.

The vector triple product, expression 28, leads to a vector that lies in the same plane as **A** and **B**. It may be expanded as follows:

$$(A \times B) \times C = B(C \cdot A) - A(B \cdot C) \qquad [31]$$

Proof of this expansion will not be given, but it is quite simply obtained by expansion of both sides of equation 31 in rectangular coordinates.

Vector Fields. Vector and scalar fields may be plotted or diagrammed in various ways, and plotting such fields is helpful in understanding their mathematical behavior.

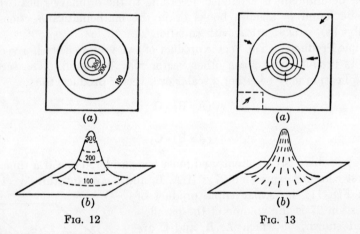

Fig. 12   Fig. 13

Contour maps are particularly interesting because they are plots of a scalar field, elevation. Figure 12, for instance, is a map of a mountain. It is a singularly symmetrical mountain, rising to a peak in the center of the map. Elevation, as plotted on a contour map, is a scalar quantity: each point is at an elevation of so many feet, and, when all points of equal elevation are connected by contour lines, the form of the earth's surface is completely defined.

Consider a marble placed upon the mountain of Fig. 12; it will try to roll down hill. Wherever it is placed upon the mountain slope, a certain force will be required to hold it, and in this way a vector field of force is defined. In Fig. 13a, another map of the mountain, arrows

GRADIENT

are drawn to indicate the amount and direction of the force required to hold the marble at various places. Such arrows are not essentially different from the *hachure* markings commonly used on geographic maps to indicate mountains, as shown in Fig. 13b. A map with hachure marks may be considered to be a rather primitive plot of a vector field.

It is apparent that there is a relation between the scalar field of elevation and the vector field of force-on-a-marble. It is a simple and familiar one: The force is dependent upon the steepness of the slope, or, in other words, upon the rate of change of elevation with respect to distance.

This rate of change is a derivative, similar in nature to the ordinary derivative of differential calculus. It is complicated, however, by the necessity of finding the direction of steepest slope to determine the *direction* in which the marble will tend to roll. The steepest slope at a given point is known as the **gradient** at that point. It is a vector quantity at each point, and therefore constitutes a vector field.

Gradient. In the lower left corner of Fig. 13a, a section of the map is indicated by dash lines. This rectangular section is enlarged in Fig. 14, and a pair of coordinate axes is superimposed for reference purposes. The gradient in this small section of the field is practically uniform, being much the same at all points in magnitude and direction, and it is indicated by a vector. This vector of gradient is made up of two components, one the steepness in the x direction, $\partial P/\partial x$, and the other the steepness in the y direction, $\partial P/\partial y$.

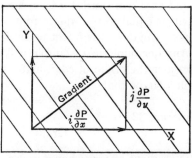

Fig. 14

Elevation is indicated by the symbol [3] P, and in this two-dimensional field

$$\text{Gradient} = i\,\frac{\partial P}{\partial x} + j\,\frac{\partial P}{\partial y} \qquad [32]$$

Two characteristics of gradient are so important that they must be mentioned at once. First, the gradient vector will always be at right angles to the contour lines. This is evident because the gradient is the steepest slope; the steepest slope will be found in descending unit elevation in the shortest horizontal distance; travel from one contour line to the next results in unit change of elevation and this is accomplished

---

[3] The symbol P is used because elevation is a gravitational *potential*.

in the shortest horizontal distance by taking a path perpendicular to the contour lines.

Second, the closer the contour lines are spaced, the steeper the slope and the greater the gradient.

If the elevation at every point is known and can be expressed analytically, giving P in terms of x and y, it is easy to apply equation 32. A simple example may be considered, referring to Fig. 14: Elevation at the origin is 1000 feet, and the hillside slopes up to the northeast. When traveling due east the ground rises 4 feet per mile, while toward the north the slope is 3 feet per mile, and so (within this limited region) elevation at any point can be found from

$$P = 1000 + 4x + 3y \qquad [33]$$

Substituting equation 33 into equation 32 gives

$$\text{Gradient} = \mathbf{i}4 + \mathbf{j}3 \qquad [34]$$

This indicates that the slope is everywhere the same (x and y do not appear in equation 34), and in such a direction that a rolling marble would go 3 feet south for each 4 feet west. The steepest slope, or gradient, is equal to the square root of the sum of the squares of the component slopes and is 5 feet per mile.

Another numerical example will illustrate a slightly less simple case: Consider the origin of coordinates to be at the top of a 1000-foot hill of such a shape that

$$P = 1000 - x^2 - y^2 \qquad [35]$$

Equation 32 gives

$$\text{Gradient} = -\mathbf{i}2x - \mathbf{j}2y \qquad [36]$$

The gradient is zero when x and y are zero, which is natural, for that is the precise top of the hill. As one moves out in any direction, the gradient becomes greater. The gradient, moreover, differs in direction at different points of the hill's surface. Full information is given by equation 36.

The above discussion of gradient refers to a two-dimensional field or surface. An electric field is a three-dimensional field in space. Flow of water as, for example, circulating currents in a large tank, may be represented by a three-dimensional vector field wherein the vectors represent velocity of flow. Temperature in a large block of unequally heated metal is a three-dimensional scalar field. Flow of heat in such a block of metal is determined by the temperature gradient.

A three-dimensional gradient, such as this gradient of temperature, is exactly analogous to the two-dimensional gradient of elevation that

GRADIENT

has been considered, and its definition is similar. Given a three-dimensional scalar field P, the gradient of P is the vector field given by

$$\text{Gradient of } P = \mathbf{i}\frac{\partial P}{\partial x} + \mathbf{j}\frac{\partial P}{\partial y} + \mathbf{k}\frac{\partial P}{\partial z} \qquad [37]$$

Equation 32, in which P is a function of x and y only, is a special case of this more general definition.

Certain statements have been made about gradient that have not been proved. They have merely been illustrated by numerical examples. For proof, consider an infinitesimal length $d\mathbf{s}$, the components of which are

$$d\mathbf{s} = \mathbf{i}\,dx + \mathbf{j}\,dy + \mathbf{k}\,dz \qquad [38]$$

This short length lies in a scalar field P, of which the gradient is a vector field *defined* by equation 37. Multiplying equation 37 by equation 38 gives

$$(\text{Gradient of } P) \cdot d\mathbf{s} = \frac{\partial P}{\partial x}dx + \frac{\partial P}{\partial y}dy + \frac{\partial P}{\partial z}dz \qquad [39]$$

The right-hand member will be recognized as the total differential of P, so

$$(\text{Gradient of } P) \cdot d\mathbf{s} = dP \qquad [40]$$

As discussed in any differential calculus book, the total differential dP is the change in P as one moves a distance $d\mathbf{s}$. If one can move a distance $d\mathbf{s}$ with no change in P, so that $dP/d\mathbf{s} = 0$, the movement is said to be along an equipotential surface. By equation 14, the left-hand member of equation 40 is zero when the vector $d\mathbf{s}$ is normal to the vector (**Gradient of** P); this makes $dP = 0$, and, hence, when $d\mathbf{s}$ is normal to the vector (**Gradient of** P), $d\mathbf{s}$ lies in an equipotential surface. The vector (**Gradient of** P) is therefore normal to the equipotential surface.

Also by equation 14, the left-hand member of equation 40 is maximum if the vector $d\mathbf{s}$ is in the same direction as the vector (**Gradient of** P). Hence the direction of the gradient is the direction of the maximum value of $dP/d\mathbf{s}$, the direction of greatest rate of change of the function.

Finally, the magnitude of the gradient may be found as the square root of the sum of the squares of its Cartesian components. In this it is just like any vector, and, considering that the gradient is defined as the vector of equation 37, it will be seen that

$$\text{Magnitude of gradient of } P = \sqrt{\left(\frac{\partial P}{\partial x}\right)^2 + \left(\frac{\partial P}{\partial y}\right)^2 + \left(\frac{\partial P}{\partial z}\right)^2} \qquad [41]$$

Divergence. The quantity called gradient is a rate of change in a scalar field. A vector field also changes from point to point, but in a more complicated manner. It cannot be said to have a gradient, but there are other ways in which the rate of change of a vector field can be described. One of the most useful is known as **divergence**. The divergence of a vector field **A** is:

$$\text{Divergence of } \mathbf{A} = \frac{\partial A_x}{\partial x} + \frac{\partial A_y}{\partial y} + \frac{\partial A_z}{\partial z} \qquad [42]$$

A_x is the magnitude of the x component of **A**, and, since **A** is a three-dimensional vector field, A_x is a three-dimensional scalar field. A_x

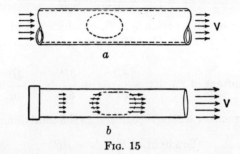

Fig. 15

may vary from point to point and is, in general, a function of x, y, and z. Its derivative with respect to x is the first term of equation 42. The second and third terms are found in similar manner from the y and z components of **A**.

It has been mentioned that divergence is a useful quantity. Its application is nicely illustrated in the flow of fluids. Figure 15a shows a pipe through which water is flowing. The dash line within the pipe represents an imaginary surface; water passes through this surface. The surface is completely closed, and water will flow in through one side and out through the other. The water may flow in any irregular fashion whatever, but (since water is incompressible) the same amount of water must flow out that flows in. It will be proved a little later that this is the same as saying that water, being incompressible, must flow in such a way that, if its velocity is represented by the vector field **V**, the divergence of **V** must everywhere be zero. It is from this concept that the name "divergence" arises: Water cannot diverge from any point for it would leave a vacuum; it cannot converge to any point for it is incompressible.

But the flow of air is different. Figure 15b represents a tube of compressed air, capped on one end. A similar cap has just been removed from the other end and air is rushing out. Consider the closed surface within the tube that is represented by a dash line; because the air is expanding, more air is passing out (through one end) of the indicated surface than is entering the surface (through the other end). Consequently there is a divergence of air. There is divergence at every point where air is expanding and, if velocity of air is the vector field **V**, the divergence of **V** is not zero.

Curl. Another important way of describing a rate of change in a vector field is given the name of **curl**. Consider a tub of water; Fig.

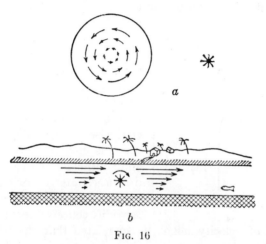

Fig. 16

16a shows the tub as seen from above. The water in it has been stirred with a paddle and the vectors represent velocity **V**. A small paddle-wheel is shown beside the tub; if this paddle-wheel, mounted on frictionless bearings, is dipped into the center of the tub, it will be turned in a counterclockwise direction. At whatever point the paddle-wheel may be place in the tub, it will be turned by the water, for, even if it is not in the center of the tub, the water will be going more rapidly past one side of the wheel than past the other. The turning of the paddle-wheel is an indication that the water is moving in the tub in such a way that the vector field of velocity has a rate of change of the type called "curl."

The name "curl" indicates an association with motion in curved lines. This is not necessary, however, for straight-line motion may also have curl. If water flows in a canal, as in Fig. 16b, in such a way that it flows more rapidly near the surface than it does along the bottom of the canal, every particle of water may move in a straight line, but nevertheless there is curl, as will be recognized when an exploring paddle-wheel is considered. The exploring wheel as seen in the figure will be turned in a clockwise direction, for the stream is more rapid on its upper blades than on its lower ones.

Figure 17 shows a map of another canal, in which the water flows without curl. In the straight part of the canal, the velocity of flow is uniform, and it is obvious that the paddle-wheel at position a will not turn. At b, in a bend of the canal, it is possible for water to turn the corner without curl, provided it flows faster along the inner margin of the channel in just the right proportion. An enlarged view of the paddle-wheel at b is shown (it must be understood that the exploring paddle-wheel is in fact so small that it does not interfere with the flow of water), and little arrows indicate the reaction of the water on each of its blades. Because of the curvature of the lines of flow, more than half of the blades are driven clockwise. But the velocity of water is greater on the inner side, and, although fewer blades are driven counterclockwise, they are each acted on more forcefully. It is readily conceivable that curvature and variation of velocity might be so related that the wheel would have no total tendency to turn. Curved motion is therefore possible without curl. This kind of flow is, as a matter of fact, characteristic of a truly frictionless liquid. It is the purpose of "streamlining" to provide a surface past which air or water will flow with a minimum of curl, for motion with curl develops eddies that waste energy.

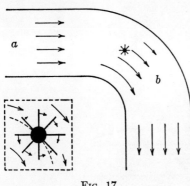

Fig. 17

Divergence of a vector field is a *scalar* quantity. There is divergence from a point or to a point (positive or negative), but no idea of direction is involved. Curl of a vector field, on the contrary, is a *vector*. If curl is visualized as an eddy, it is evident that the eddy must be about some axis—perhaps a vertical axis, perhaps horizontal, perhaps at some angle. The direction of such an axis is, by definition, the direction of the vector that represents curl. Referring to the hypothetical paddle-wheel, when it is in the position in which it turns most rapidly, its axis is in the direction of the curl vector. Each component of the vector of curl may be found by placing the paddle-wheel axis parallel to the appropriate axis of coordinates.

The sense of the curl vector is determined by the direction of rotation of the paddle-wheel: if the paddle-wheel turns a right-hand screw, it will screw itself in the direction of the curl vector, as in Fig. 18.

Mathematically the curl of a vector field **A** is defined by

$$\text{Curl of } \mathbf{A} = \mathbf{i}\left(\frac{\partial A_z}{\partial y} - \frac{\partial A_y}{\partial z}\right) + \mathbf{j}\left(\frac{\partial A_x}{\partial z} - \frac{\partial A_z}{\partial x}\right) + \mathbf{k}\left(\frac{\partial A_y}{\partial x} - \frac{\partial A_x}{\partial y}\right) \quad [43]$$

It will be shown later that this is equivalent to the physical concept of curl that has been discussed in the preceding paragraphs.

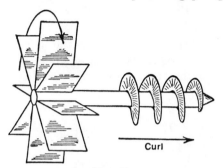

FIG. 18

Nabla. Time is saved in writing the equations of vector analysis, and, what is more important, they are made easier to remember, by the use of a symbol known as "nabla" and written ∇.[4] Its formal definition is

$$\nabla \equiv \mathbf{i}\frac{\partial}{\partial x} + \mathbf{j}\frac{\partial}{\partial y} + \mathbf{k}\frac{\partial}{\partial z} \quad [44]$$

It will be seen that this symbol by itself has no meaning. It has the formal appearance of a vector for which the x, y, and z components are, respectively, $\partial/\partial x$, $\partial/\partial y$, and $\partial/\partial z$. But, like $\partial/\partial x$, *nabla* is an *operator* and must have something on which to operate. If it is allowed to operate on a scalar function, it gives the gradient of that function; operating on P it gives

$$\nabla P = \mathbf{i}\frac{\partial}{\partial x}P + \mathbf{j}\frac{\partial}{\partial y}P + \mathbf{k}\frac{\partial}{\partial z}P \quad [45]$$

and this, by comparison with equation 37, is seen to be the gradient of P. The expression ∇P is expanded exactly as if it were the product of a vector quantity ∇ by a scalar quantity P. Actually it is not a multiplication at all, but an operation of differentiation.

[4] This symbol is frequently given the name "del," but this has been found to lead to confusion with the Greek letter "delta," which is similar in appearance but which conveys an entirely different mathematical meaning. Therefore the Hamiltonian name of "nabla" is returning to common use and is recommended, for example, by the American Institute of Electrical Engineers. The original nabla was a Hebrew harp of triangular shape, ל ב נ, the psaltery of the Psalms.

The symbol ∇, as defined in equation 44, can be put through many algebraic transformations as if it were indeed a vector quantity, and this is one of the advantages of its use. For instance, a quantity written formally as the dot product of ∇ and a vector field gives the divergence of that field. Referring to equation 22 we write:

$$\nabla \cdot \mathbf{B} = \frac{\partial}{\partial x} B_x + \frac{\partial}{\partial y} B_y + \frac{\partial}{\partial z} B_z \qquad [46]$$

which, by comparison with equation 42, is the divergence of **B**.

Similarly, a quantity written as the cross product of ∇ and a vector field gives the curl of that field. Referring to equation 24, the cross product is expanded:

$$\nabla \times \mathbf{B} = \mathbf{i}\left(\frac{\partial}{\partial y} B_z - \frac{\partial}{\partial z} B_y\right) + \mathbf{j}\left(\frac{\partial}{\partial z} B_x - \frac{\partial}{\partial x} B_z\right)$$

$$+ \mathbf{k}\left(\frac{\partial}{\partial x} B_y - \frac{\partial}{\partial y} B_x\right) \qquad [47]$$

and by comparison with equation 43 this is recognized as the curl of **B**.

The equation for curl is formally similar to the equation for the cross product, and it also may be expressed as a determinant. The determinantal form is much easier to remember than the expanded form of equation 47: it is

$$\nabla \times \mathbf{B} = \begin{vmatrix} \mathbf{i} & \mathbf{j} & \mathbf{k} \\ \dfrac{\partial}{\partial x} & \dfrac{\partial}{\partial y} & \dfrac{\partial}{\partial z} \\ B_x & B_y & B_z \end{vmatrix} \qquad [48]$$

The following tabulation collects information relating to differential operations on vector and scalar fields.

Type of Operation	Symbol	Must Be Applied to:	Yields:
Gradient of A	∇A	Scalar field	Vector field
Divergence of **A**	$\nabla \cdot \mathbf{A}$	Vector field	Scalar field
Curl of **A**	$\nabla \times \mathbf{A}$	Vector field	Vector field

Divergence and *curl* are both derivative operations on vector fields. They are both space derivatives; that is, they are partial derivatives with respect to the distances x, y, and z. The essential mathematical difference is quite simple. *Divergence is a rate of change of the field strength in the direction of the vector field*; thus in the definition of divergence, equation 46, one takes the partial derivative with respect to x of the x component of the field, the partial derivative with respect to y of the y component, and with respect to z of the z component. A physical example is seen in Fig. 15b. *Curl is a rate of change of the field strength in a direction at right angles to the field*; thus, in equation 47 one finds curl by taking the y and z derivatives of the x component, the x and z derivatives of the y component, and so on, but not the x derivative of the x component. Figure 16b shows a simple example of a field that varies across the field but not along the field, and so does Fig. 19b.

Illustrative Examples. The most satisfactory way to become familiar with gradient, divergence, and curl is by study of a few simple illustrations of vector fields. These will be mere geometrical fields, with no physical meaning attached at the present time. Some of them will later be found to be of electromagnetic importance.

Only two-dimensional fields will be considered in this section; an extension to three dimensions is simple when the two-dimensional case is understood, and two-dimensional illustrations are clearer because they are less obscured by mathematical manipulation. Moreover, a great many practical cases in three dimensions can be reduced to two-variable problems by such choice of coordinate axes that the quantities being studied are functions of two variables only.

When working with a scalar field that is the same for all values of z and therefore has derivatives with respect to x and y only, it is evident from equation 45 that the gradient is merely

$$\nabla P = \mathbf{i}\frac{\partial P}{\partial x} + \mathbf{j}\frac{\partial P}{\partial y} \qquad [49]$$

For a two-variable vector field **A**, divergence is

$$\nabla \cdot \mathbf{A} = \frac{\partial A_x}{\partial x} + \frac{\partial A_y}{\partial y} \qquad [50]$$

If it is also true that the vector field has no component in the z direction ($A_z = 0$), the curl is

$$\nabla \times \mathbf{A} = \mathbf{k}\left(\frac{\partial A_y}{\partial x} - \frac{\partial A_x}{\partial y}\right) \qquad [51]$$

Equations of vector analysis are collected for ready reference in Table II (inside back cover).

Example 1. Consider a vector field defined by the equations

$$A_x = 1 \qquad A_y = 2 \qquad [52]$$

These are, of course, the components of the vector field **A** and

$$\mathbf{A} = \mathbf{i}A_x + \mathbf{j}A_y = \mathbf{i}1 + \mathbf{j}2 \qquad [53]$$

In Fig. 19a this field is indicated, the lines showing the direction of the vectors; it has already been mentioned that graphical representation of a vector field is not easy and some effort of visualization will be required. From the defining equation it is seen that the field intensity at any point (P in the figure) is 1 unit in the x direction and 2 units in the y direction, as shown. This is the same everywhere, for neither A_x nor A_y is a

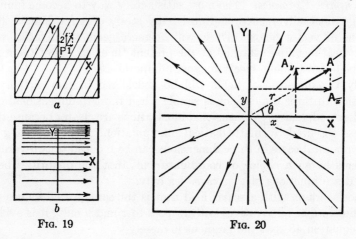

Fig. 19   Fig. 20

function of either x or y. This field has neither divergence nor curl. That there is no divergence is shown when A_x and A_y are substituted into equation 50, for the partial derivatives are both zero. Likewise there is no curl, for the partial derivatives of equation 51 are both zero.

Example 2. Consider a vector field defined by

$$A_x = y + 10 \qquad A_y = 0 \qquad [54]$$

This field is shown in Fig. 19b. The field is entirely in the x direction because A_y is everywhere zero, and it becomes more intense toward the top because A_x increases with y; this is shown in the figure by drawing the lines that indicate direction of the field closer together in the more intense region (as is customarily done with magnetic and electric lines of force). If y were less than -10 the direction of the field would reverse, and along the line $y = -10$ there is zero field, but the diagram is not extended to that region.

Equation 50 tells us that this field has no divergence. In applying equation 51, we find that $\partial A_y/\partial x = 0$, but $\partial A_x/\partial y = 1$. Hence the curl of the field is everywhere $-\mathbf{k}$. (Consider the exploring paddle-wheel or curl-meter of Fig. 18.)

Example 3. Next it is desired to consider a field that has everywhere unit intensity and is everywhere radial from the origin of coordinates in direction. Such a field is suggested in Fig. 20, in which an attempt has been made to keep the density of radial lines everywhere the same. Consider any point p with coordinates x and y; since $\mathbf{A}$ is to be radial, A_x must be $A \cos \theta$ and A_y must be $A \sin \theta$. Hence if A is everywhere unity,

$$A_x = \frac{x}{\sqrt{x^2 + y^2}} \qquad A_y = \frac{y}{\sqrt{x^2 + y^2}} \qquad [55]$$

These equations define the field at all points in any quadrant. In computing divergence and curl it is necessary to determine the four partial derivatives:

$$\frac{\partial A_x}{\partial x} = \frac{1}{(x^2 + y^2)^{1/2}} - \frac{x^2}{(x^2 + y^2)^{3/2}} \qquad [56]$$

$$\frac{\partial A_y}{\partial y} = \frac{1}{(x^2 + y^2)^{1/2}} - \frac{y^2}{(x^2 + y^2)^{3/2}} \qquad [57]$$

$$\frac{\partial A_y}{\partial x} = -\frac{xy}{(x^2 + y^2)^{3/2}} \qquad [58]$$

$$\frac{\partial A_x}{\partial y} = -\frac{xy}{(x^2 + y^2)^{3/2}} \qquad [59]$$

Adding the first two to obtain divergence gives

$$\nabla \cdot \mathbf{A} = \frac{1}{(x^2 + y^2)^{1/2}} = \frac{1}{r} \qquad [60]$$

Subtracting the fourth from the third to obtain curl gives

$$\nabla \times \mathbf{A} = 0 \qquad [61]$$

The divergence in this field is particularly interesting. The presence of divergence is associated with the necessity for starting new radial lines in the diagram in order to indicate a constant intensity of field. It is apparent that, if lines start within a region of space, there must be more lines coming out of that region than enter it; in such a case there is divergence in that region. Qualitatively, it may be seen in Fig. 20 that divergence is greatest near the origin, for that is where most lines originate; quantitatively, equation 60 tells us that divergence is inversely proportional to radius and increases without limit as the origin is approached.

A differentiating operation upon a field can be repeated, giving a quantity analogous to a second derivative. In the above example the divergence of **A** is itself a scalar field of which the gradient can be found. This will give the gradient of the divergence of **A** [symbolically $\nabla(\nabla \cdot \mathbf{A})$].

$$\nabla(\nabla \cdot \mathbf{A}) = \nabla \frac{1}{\sqrt{x^2+y^2}} = \mathbf{i}\frac{\partial}{\partial x}(x^2+y^2)^{-1/2} + \mathbf{j}\frac{\partial}{\partial y}(x^2+y^2)^{-1/2}$$

$$= -\mathbf{i}\frac{x}{r^3} - \mathbf{j}\frac{y}{r^3} \qquad [62]$$

This result is a vector field, which we can again differentiate to find its curl or its divergence. Finding its curl, two partial derivatives are needed:

$$\frac{\partial}{\partial x}\left(-\frac{y}{r^3}\right) = \frac{\partial}{\partial x}[-y(x^2+y^2)^{-3/2}] = (-y)(-\tfrac{3}{2})(x^2+y^2)^{-5/2}(2x)$$

$$\frac{\partial}{\partial y}\left(-\frac{x}{r^3}\right) = \frac{\partial}{\partial y}[-x(x^2+y^2)^{-3/2}] = (-x)(-\tfrac{3}{2})(x^2+y^2)^{-5/2}(2y)$$

[63]

Subtracting these partial derivatives gives the curl, and since they are equal the curl is zero. That is, the curl of the gradient of $1/r$ is zero:

$$\nabla \times \nabla \frac{1}{r} = 0 \qquad [64]$$

It is not a coincidence or a special property of this particular field that the curl of the gradient is zero; it will be shown in the next paragraph that the curl of the gradient of any scalar field is always identically zero.

A Field That Is the Gradient of Something Has No Curl. ($\nabla \times \nabla F \equiv 0$.) To prove this theorem, consider any scalar field. This field will be denoted by F. First write its gradient, using equation 45:

$$\nabla F = \mathbf{i}\frac{\partial F}{\partial x} + \mathbf{j}\frac{\partial F}{\partial y} + \mathbf{k}\frac{\partial F}{\partial z} \qquad [65]$$

Note that this gradient is a vector with components $\partial F/\partial x$, $\partial F/\partial y$, and $\partial F/\partial z$. The curl of this vector is found by substituting into equation 47:

$$\nabla \times (\nabla F) = \mathbf{i}\left(\frac{\partial}{\partial y}\frac{\partial F}{\partial z} - \frac{\partial}{\partial z}\frac{\partial F}{\partial y}\right) + \mathbf{j}\left(\frac{\partial}{\partial z}\frac{\partial F}{\partial x} - \frac{\partial}{\partial x}\frac{\partial F}{\partial z}\right)$$

$$+ \mathbf{k}\left(\frac{\partial}{\partial x}\frac{\partial F}{\partial y} - \frac{\partial}{\partial y}\frac{\partial F}{\partial x}\right) \qquad [66]$$

THE CURL OF THE GRADIENT IS ZERO

This completes the proof, for, since the order of differentiation of a second partial derivative is immaterial, each of the parentheses in the above expression is identically zero.

Another similar theorem states that for any vector field **A**, $\nabla \cdot \nabla \times \mathbf{A} \equiv 0$. In words, *a field that is the curl of something has no divergence*. This is illustrated by Problems 11 and 12. The general theorem is easily proved (as Problem 13) by a method quite similar to the proof of the identity $\nabla \times \nabla F \equiv 0$. As in that case, expansion leads to equal and opposite second partial derivatives.

There is one other repeated differentiation of great importance in applications to physical problems, that is, the divergence of the gradient. The divergence of the gradient is of so much importance that it is given a special name; it is called the "Laplacian," after the famous French mathematician of a century and a half ago. It is, moreover, given a special symbol; although the Laplacian of F would properly be written $\nabla \cdot \nabla F$, it has become customary to abbreviate this to $\nabla^2 F$, the meaning of course being the same. The Laplacian, being divergence, is a scalar field. Expressed in terms of second partial derivatives, it is extraordinarily simple and of obvious importance:

$$\nabla^2 F = \nabla \cdot (\nabla F) = \nabla \cdot \left(\mathbf{i} \frac{\partial F}{\partial x} + \mathbf{j} \frac{\partial F}{\partial y} + \mathbf{k} \frac{\partial F}{\partial z} \right)$$

$$= \frac{\partial}{\partial x} \frac{\partial F}{\partial x} + \frac{\partial}{\partial y} \frac{\partial F}{\partial y} + \frac{\partial}{\partial z} \frac{\partial F}{\partial z}$$

$$= \left(\frac{\partial^2}{\partial x^2} + \frac{\partial^2}{\partial y^2} + \frac{\partial^2}{\partial z^2} \right) F \qquad [67]$$

The Laplacian is not, in general, equal to zero. (The curl of the gradient and the divergence of the curl are the only two second-derivative operations that are always identically zero.) Yet the Laplacian is frequently zero in physical problems, depending upon the physical conditions. In electrostatics, for instance, the Laplacian of the electric potential is zero in any space that does not contain electric charge; this will be shown in a later chapter.

Equation 67 gives the Laplacian of a scalar field F. The Laplacian of a vector field is also useful. The Laplacian of the vector field **A** is written $\nabla^2 \mathbf{A}$, and it is interpreted to mean

$$\nabla^2 \mathbf{A} = \nabla^2 (\mathbf{i} A_x + \mathbf{j} A_y + \mathbf{k} A_z) \qquad [68]$$

The Laplacian of a vector field is therefore the vector sum of the Laplacians of the three scalar components of the vector field.[5] It is frequently important to know whether the Laplacian of a vector field is zero. The answer is that it is zero if and only if the Laplacians of the component scalar fields, A_x, A_y, and A_z, are each independently zero.

Polar Coordinates. It will have been noticed that several examples and problems in the earlier part of the chapter are naturally adapted to the use of polar coordinates. Example 3 and Fig. 20, for instance, have symmetry about the origin in such a way that reference to the radial distance r and the angle θ could hardly be avoided. Would it not be possible, then, to use polar coordinates instead of rectangular coordinates in connection with fields of radial symmetry, and thereby simplify the calculations?

That question introduces an aspect of vector analysis that is of the utmost importance. It is this. Vector analysis is fundamentally independent of coordinate systems. An electric field, or a field of velocity, or a field of force exists, physically, whether or not any mathematician has yet laid out a set of coordinate axes. The field has divergence or curl or gradient, or it has not, depending upon the properties of the field itself and without regard to any system of coordinates. Scalar products, vector products, sums and differences of vectors, line integrals, and surface integrals are all of significance without reference to coordinates. Nature does not provide systems of coordinates (except in special cases in which the properties of matter are different in different directions).

It is evident that a system of mathematics in which general relations can be stated without reference to coordinates is simpler than a system in which arbitrary axes must first be introduced as a frame upon which to hang the mental processes. Hence the theorems and generalizations of vector analysis are much easier to comprehend than the similar statements of coordinate geometry.

For actual computation, unfortunately, it is usually necessary to refer to coordinates. The axes are required for calculations, for how can a field be defined at every point in space except by identifying each point by means of coordinates? But, at least, generalized thinking can be done in terms of vector analysis, and then any convenient set of coordinates can be used to facilitate computation. It is for this purpose that operations of vector analysis, after being defined in terms of the vectors and the fields themselves (as in equations 14 and 19), are also expressed with reference to a coordinate system (as in equations 22 and 24, and 45, 46, and 47).

---

[5] This may be taken as the definition of the Laplacian of a vector field. An alternative definition that gives the Laplacian the same value but avoids defining it in terms of specific coordinates is: $\nabla^2 \mathbf{A} = \nabla(\nabla \cdot \mathbf{A}) - \nabla \times (\nabla \times \mathbf{A})$.

POLAR COORDINATES

Some coordinate systems are more convenient than others in specific computations. As illustrated in Problem 6, the choice of coordinate system makes no difference in the result. And, since polar coordinates are better adapted to many problems than are rectangular coordinates, it will be desirable to express some of the vector operations in polar coordinates.

Consider a two-dimensional polar system using as coordinates radial distance r and angle θ. First, it is necessary to be able to find r and θ when x and y are known, or vice versa. The formulas are familiar:

$$x = r \cos \theta \qquad y = r \sin \theta$$
$$r = \sqrt{x^2 + y^2} \qquad \theta = \tan^{-1} \frac{y}{x} \qquad [69]$$

Next, it is necessary to express in polar coordinates a vector that is given in rectangular coordinates. If A_x and A_y are known, it must be possible to find A_r and A_θ (A_r is defined as the length of the component of **A** in the radial direction, A_θ is the length of the component of **A** that is normal to A_r). Figure 21 shows the relation: at a point with coordinates (x,y) or (r,θ), there is a vector **A**. The polar components of the vector, A_r and A_θ, are shown; so also are the rectangular components A_x and A_y. Two right triangles are drawn; A_x is the hypotenuse of one and A_y is the hypotenuse of the other. One angle of each triangle is θ. From the construction of the figure it may be seen that

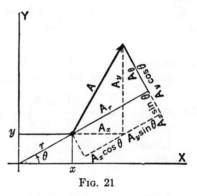

Fig. 21

$$A_r = A_x \cos \theta + A_y \sin \theta$$
$$A_\theta = A_y \cos \theta - A_x \sin \theta \qquad [70]$$

This is the desired relation when finding polar components from rectangular components; when the polar components are known and the rectangular ones are desired, the following equations are used. They are obtained from equations 70 by simultaneous solution, or by slightly different construction in Fig. 21.

$$A_x = A_r \cos \theta - A_\theta \sin \theta$$
$$A_y = A_r \sin \theta + A_\theta \cos \theta \qquad [71]$$

Gradient in polar coordinates is now easily found. In two-dimensional rectangular coordinates it is

$$\nabla P = \mathbf{i}\frac{\partial P}{\partial x} + \mathbf{j}\frac{\partial P}{\partial y} \qquad [49]$$

From equations 69

$$x = r\cos\theta \quad \text{and} \quad y = r\sin\theta$$

so, by the rules for differentiation of a composite function (see a calculus book):

$$\frac{\partial P}{\partial x} = \frac{\partial P}{\partial r}\cos\theta - \frac{\partial P}{\partial \theta}\frac{\sin\theta}{r} \qquad [72]$$

$$\frac{\partial P}{\partial y} = \frac{\partial P}{\partial r}\sin\theta + \frac{\partial P}{\partial \theta}\frac{\cos\theta}{r} \qquad [73]$$

Gradient is a vector, and it is seen from the expression for ∇P that its x and y components are given by equations 72 and 73 respectively. But it is not sufficient to know the x and y components; it is necessary to find the r and θ components. This is done by means of equations 70, substituting equation 72 for the x component and equation 73 for the y component, giving the following radial and angular components:

$$(\nabla P)_r = \left(\frac{\partial P}{\partial r}\cos\theta - \frac{\partial P}{\partial \theta}\frac{\sin\theta}{r}\right)\cos\theta + \left(\frac{\partial P}{\partial r}\sin\theta + \frac{\partial P}{\partial \theta}\frac{\cos\theta}{r}\right)\sin\theta$$

$$= \frac{\partial P}{\partial r} \qquad [74]$$

$$(\nabla P)_\theta = \left(\frac{\partial P}{\partial r}\sin\theta + \frac{\partial P}{\partial \theta}\frac{\cos\theta}{r}\right)\cos\theta - \left(\frac{\partial P}{\partial r}\cos\theta - \frac{\partial P}{\partial \theta}\frac{\sin\theta}{r}\right)\sin\theta$$

$$= \frac{1}{r}\frac{\partial P}{\partial \theta} \qquad [75]$$

Hence, if $\mathbf{1}_r$ is a radial unit vector at any point under consideration and $\mathbf{1}_\theta$ is a unit vector normal to $\mathbf{1}_r$, gradient in two-dimensional polar coordinates is

$$\nabla P = \mathbf{1}_r\frac{\partial P}{\partial r} + \mathbf{1}_\theta\frac{1}{r}\frac{\partial P}{\partial \theta} \qquad [76]$$

On page 20 the shape of a hill was described by equation 35. In polar coordinates the elevation of each point is

$$P = 1000 - r^2 \qquad [77]$$

POLAR COORDINATES

(Equation 77 may be obtained from 35 by means of 69 if the transformation is not obvious.) To find the gradient, which is the slope of the hill, use equation 76:

$$\nabla P = -1, 2r \quad [78]$$

Is this equivalent to the value of gradient given in equation 36, computed in rectangular coordinates?

Divergence can also be expressed in polar coordinates. Starting with rectangular coordinates, in two dimensions,

$$\nabla \cdot \mathbf{A} = \frac{\partial A_x}{\partial x} + \frac{\partial A_y}{\partial y} \quad [50]$$

Substituting equations 72 and 73 gives

$$\nabla \cdot \mathbf{A} = \frac{\partial A_x}{\partial r} \cos\theta - \frac{\partial A_x}{\partial \theta} \frac{\sin\theta}{r} + \frac{\partial A_y}{\partial r} \sin\theta + \frac{\partial A_y}{\partial \theta} \frac{\cos\theta}{r} \quad [79]$$

The components of **A** must now be changed from rectangular components to polar components by means of equations 71.

$$\nabla \cdot \mathbf{A} = \frac{\partial (A_r \cos\theta - A_\theta \sin\theta)}{\partial r} \cos\theta - \frac{\partial (A_r \cos\theta - A_\theta \sin\theta)}{\partial \theta} \frac{\sin\theta}{r}$$
$$+ \frac{\partial (A_r \sin\theta + A_\theta \cos\theta)}{\partial r} \sin\theta$$
$$+ \frac{\partial (A_r \sin\theta + A_\theta \cos\theta)}{\partial \theta} \frac{\cos\theta}{r} \quad [80]$$

The partial derivatives of equation 80 are now expanded, noting that θ is not a function of r but that A_r and A_θ are each functions of both r and θ. When the resulting terms are collected, the final expression for divergence is

$$\nabla \cdot \mathbf{A} = \frac{\partial A_r}{\partial r} + \frac{1}{r} \frac{\partial A_\theta}{\partial \theta} + \frac{A_r}{r} \quad [81]$$

Curl can be expressed in polar coordinates also, and the transformation from rectangular coordinates is similar to the transformation for gradient. The resulting polar expression for curl in two dimensions is

$$\nabla \times \mathbf{A} = \mathbf{k} \left(\frac{\partial A_\theta}{\partial r} - \frac{1}{r} \frac{\partial A_r}{\partial \theta} + \frac{A_\theta}{r} \right) \quad [82]$$

Example 4. To illustrate the use of equations 81 and 82, find divergence and curl of the vector field that is discussed in Example 3, page 29, and illustrated in Fig. 20. From the description of the field it is evident that its polar components are:

$$A_r = 1 \quad \text{and} \quad A_\theta = 0 \qquad [83]$$

From equation 81 the divergence of the field is $1/r$. Since each term of equation 82 is zero, the curl is zero. These are the same as the results found previously using rectangular coordinates, as given in equations 60 and 61, but the computation is so much simpler in polar coordinates that the advantage of their use is apparent.

Tabulation. Formulas for gradient, divergence, curl, and the Laplacian in rectangular, cylindrical, and spherical coordinates are included in Table II, inside the back cover.

PROBLEMS

1. Complete the tabulation of equation 21 to include all possible scalar and vector products of the unit vectors i, j, and k.

2. (a) Prove that equation 24 is correct. (b) Prove that equation 31 is correct.

3. A farm has the shape of a parallelogram, one boundary line running east 7 miles and another directly northeast 5 miles. Using equation 24, find the area of the farm.

4. Rain, blown by a south wind, falls at an angle of 30 degrees to the vertical at a speed of 60 feet per second. There is 1 ounce of rain in each cubic yard of air. How much rain falls on each square yard of the south wall of a building? On the west wall? On the flat roof? Use equation 22 for this problem, representing area by a vector.

5. Draw a contour map of the hill of equation 35 and a sketch of its shape in three dimensions. Draw vectors on the contour map to show gradient at the following points: (0,0), (0,1), (0,3), (0,−3), (3,0), (2,2), (2,1), (1,2), (−1,−2), (−2,1), (−1,2).

6. If, in Fig. 19b, the coordinate axes had been chosen at 45 degrees to their indicated position, the field would have been defined by $\mathbf{A} = \dfrac{y - x + 10\sqrt{2}}{2} (\mathbf{i} + \mathbf{j})$. Find the curl and divergence at every point using these less fortunately chosen axes.

7. $V_x = \sin y$, $V_y = 0$. Sketch the field of V (as in Fig. 19) and find its divergence and curl.

8. $V_x = \dfrac{x}{x^2 + y^2}$ $V_y = \dfrac{y}{x^2 + y^2}$. Sketch the field of V and find its divergence and curl.

9. Sketch contour lines of constant divergence for equation 60, and sketch the vector field of the gradient of this divergence, from equation 62.

10. Find the curl of the gradient of P in equation 35.

11. Find the divergence of the curl $[\nabla \cdot (\nabla \times \mathbf{V})]$ of the vector field defined in Problem 7, using equation 46.

12. $V_x = \dfrac{1}{\sqrt{x^2 + y^2}}$ $V_y = \dfrac{1}{\sqrt{x^2 + y^2}}$. Sketch the field defined by these equations and find its divergence and curl, and the divergence of the curl.

PROBLEMS

13. Prove that $\nabla \cdot \nabla \times \mathbf{V} \equiv 0$.

14. Find the Laplacian of P as given in equation 35.

15. Find the Laplacian of P if $P = -\ln r$ (r being the scalar distance of any point in a plane from a fixed point, so that by proper choice of coordinates $r^2 = x^2 + y^2$).

16. Find the Laplacian of P if $P = 1/r$ (r being the scalar distance from a fixed point) in both: (a) a two-dimensional field, and (b) a three-dimensional field.

17. Prove by means of equations 69 and 70 that equation 78 is equivalent to equation 36.

18. Working from equation 51, prove that equation 82 is correct.

19. Express the field of Problem 8 in polar coordinates and find its divergence and curl.

20. Determine whether it is advantageous to use polar coordinates in solving Problem 12.

21. In Table II (inside back cover), the scalar product $\mathbf{A} \cdot \mathbf{B}$ is expanded in rectangular coordinates. Expand $\mathbf{A} \cdot \mathbf{B}$ in cylindrical coordinates. Expand $\mathbf{A} \cdot \mathbf{B}$ in spherical coordinates.

22. In Table II (inside back cover), the vector product $\mathbf{A} \times \mathbf{B}$ is expanded in rectangular coordinates and expressed in the form of a determinant. Express $\mathbf{A} \times \mathbf{B}$ similarly in cylindrical and spherical coordinates.

23. Expand the determinant for curl in spherical coordinates given in Table II (inside back cover); perform as many of the indicated differentiations of products as possible and simplify the result.

$$\nabla^2 P = \left(\frac{\partial^2}{\partial x^2} + \frac{\partial^2}{\partial y^2} + \frac{\partial^2}{\partial z^2}\right) P \ldots - x^2 - y^2)$$

$x = \frac{y}{(x^2+y^2)^{3/2}} V_y = \frac{-x}{(x^2+y^2)^{3/2}}$ gives field which varies $\frac{1}{R^2}$, has 0 div

curl $= \frac{1}{R^{3/2}}$

CHAPTER III

Certain Theorems Relating to Fields

Divergence. The general idea of divergence was introduced in the previous chapter. Divergence occurs in a region in which lines of flow (the literal meaning of "flux lines") appear to originate. An equation for computing divergence (equation 46) was given, but no proof or demonstration was included to show that this equation was truly related to the physical idea of divergence. Such demonstration will now be

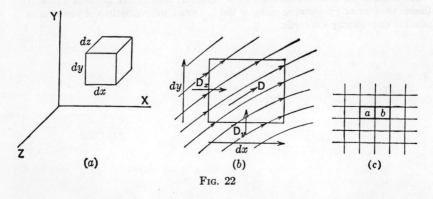

Fig. 22

given. It will not be rigorous, but it will indicate the principal steps of a rigorous proof.

Consider a small rectangular prism with its edges parallel to three coordinate axes X, Y, and Z, as in Fig. 22a. The limiting case is to be considered, in which the prism is so small that its edges are dx, dy, and dz in length. Figure 22b shows a side view of this prism, with the plane of the figure parallel to the X-Y plane. We are looking upon a side with area $dx\,dy$. Each end has area $dy\,dz$, and the top and bottom $dx\,dz$.

This small prismatic volume is located in a vector field which, for convenience, we will call **D**. Flux lines of this field pass through the prism, entering through one surface and leaving through another. We wish to find how many lines, if any, originate within the volume.

38

GAUSS'S THEOREM

Referring to Fig. 22b, the number of flux lines entering the left-hand side of the prism is equal to the area of the left-hand surface times the normal component of field strength, which is $D_x\,dy\,dz$. The number leaving the right-hand surface is different if D_x changes in the distance dx. If D_x is changing at the rate $\dfrac{\partial D_x}{\partial x}$ as one passes from left to right, the amount of change in the distance dx is $\dfrac{\partial D_x}{\partial x}\,dx$. Hence the number of flux lines leaving the right-hand surface is $\left(D_x + \dfrac{\partial D_x}{\partial x}\,dx\right)dy\,dz$. Subtracting, the number of lines that leave the right-hand side in excess of the number that enter the left-hand side is $\dfrac{\partial D_x}{\partial x}\,dx\,dy\,dz$.

Similarly, the number of lines leaving the top of the prism in excess of those entering the bottom is $\dfrac{\partial D_y}{\partial y}\,dy\,dx\,dz$; and the number leaving the front surface is greater than the number entering the back by $\dfrac{\partial D_z}{\partial z}\,dz\,dx\,dy$.

Combining these quantities, the total number of flux lines leaving the volume that do not enter it is

$$\left(\frac{\partial D_x}{\partial x} + \frac{\partial D_y}{\partial y} + \frac{\partial D_z}{\partial z}\right)dx\,dy\,dz \qquad [84]$$

But divergence is defined as the number of flux lines originating per unit volume; so, if the volume of the prism is dv,

$$\nabla \cdot \mathbf{D} = \left(\frac{\partial D_x}{\partial x} + \frac{\partial D_y}{\partial y} + \frac{\partial D_z}{\partial z}\right)\frac{dx\,dy\,dz}{dv} \qquad [85]$$

Since the volume of the prism dv is equal to $dx\,dy\,dz$, it follows that

$$\nabla \cdot \mathbf{D} = \frac{\partial D_x}{\partial x} + \frac{\partial D_y}{\partial y} + \frac{\partial D_z}{\partial z} \qquad [86]$$

and this is equation 46.

Gauss's Theorem. Now consider that space is divided into an unlimited number of small cells of volume dv, as in Fig. 22c. The number of flux lines leaving one such cell, marked a in the figure, is greater than the number entering that cell by $\nabla \cdot \mathbf{D}\,dv$. The number originating within the adjoining cell b is likewise the divergence at that location times the volume of that cell. The number of lines emanating from the two cells together, considered as a unit, is the sum of the two products of divergence and volume. Adding more cells to the group thus begun, the number of lines of flux issuing from any volume is greater than the number entering that volume by the summation (or integral) of all the individual products of divergence and volume. Hence

$$\text{Excess outward flux} = \int \nabla \cdot \mathbf{D}\, dv \qquad [87]$$

In Chapter I, flux of the vector field **D** passing through an area **a** was defined as

$$\int \mathbf{D} \cdot d\mathbf{a} \qquad [11]$$

and from this it follows that the net flux passing outward through any closed surface (the excess of the outward flux over the inward flux) is found by integrating over the whole closed surface:[1]

$$\oint \mathbf{D} \cdot d\mathbf{a} \qquad [88]$$

Now equation 88 and equation 87 are different expressions for the same quantity of flux and hence may be equated, giving

$$\oint \mathbf{D} \cdot d\mathbf{a} = \int \nabla \cdot \mathbf{D}\, dv \qquad [89]$$

This is a theorem of great importance. It relates the integral of divergence within any volume to the integral of the vector field strength over the surface enclosing that volume. It is sometimes called the divergence theorem and sometimes *Gauss's theorem* or *Green's theorem*.

A number of illustrations of the application of this theorem are given in the preceding chapter, especially with reference to Fig. 15.

---

[1] Two comments regarding notation: A small circle superimposed on the integral sign indicates integration over a closed path, either a closed line or a closed surface, depending upon whether the integration is with respect to distance or area, as indicated by the nature of the differential quantity. In this case $d\mathbf{a}$ indicates that the integration is over an area, and the circle upon the integral sign indicates that the surface over which the integral is taken must be a closed surface.

In equation 4 the differential quantity $d\mathbf{s}$, s being distance, indicates integration along a line; that it must be a closed line is shown by the circle on the integral sign.

The area $d\mathbf{a}$ is a vector quantity, as discussed in Chapter II. The direction of the vector $d\mathbf{a}$ is normal to the area and, *by convention*, it is *outward*. It is obvious that this convention is useful only for a closed surface.

Curl. The idea of curl was developed in Chapter II with reference to a hypothetical paddle-wheel in a vector field of fluid velocity. Rotation of the paddle-wheel occurred when the summation of the components of field strength tangential to the paddle-wheel's circumference failed to add to zero. This was illustrated by reference to the force exerted on specific paddles, a painfully crude illustration. The idea of curl is much better formulated in terms of the *circulation* of the vector field about the periphery of the paddle-wheel. The paddle-wheel may then be removed entirely, leaving curl defined in terms of the circulation about a small closed path.

The mathematical quantity **circulation** is the line-integral of a vector field along a given path. If the vector field is **E**, its circulation about a closed path is

$$\oint \mathbf{E} \cdot d\mathbf{s} \qquad [90]$$

This is merely a mathematical expression of the ordinary concept of circulation, as of air or water, and it is clearly the circulation of fluid about its periphery that makes a paddle-wheel turn.

Curl is a microscopic circulation. Consider the exploring paddle-wheel in a vector field, and orient it so that its speed of rotation is maximum. (This determines the orientation in which the circulation around its circumference is maximum.) Now allow the paddle-wheel to vanish, but retain its circumference as a circular path in space. The circulation of the vector field about this path (found from the above definition) depends upon the area enclosed within the path. Dividing circulation by area gives a ratio that is substantially independent of the size or shape of the path provided the path is small. This ratio is

$$\frac{\text{Circulation about a small closed path}}{\text{Area of surface bounded by that path}} \qquad [91]$$

The limit approached by this ratio as the path is allowed to shrink to a mere point is the **curl** at that point.

Curl is thus the limiting value of *circulation per unit area*. It follows that the circulation of a vector field around a closed path of infinitesimal size depends upon the curl of the field at that point and the infinitesimal area within the path. But the circulation about a small closed path also depends upon its orientation. Curl must be treated as a vector quantity. The direction of the curl vector is defined as normal to the plane in which circulation is maximum, and the circulation about an infinitesimal closed path in either that or any other plane is the scalar product of the curl vector and the area vector: $(\nabla \times \mathbf{E}) \cdot d\mathbf{a}$.

Now curl has been defined in terms of circulation. To find an expression for curl in terms of the vector field itself, we determine the circulation about a small closed path in the vector field. It is not necessary that such a path be circular. Let us assume a small rectangular closed path, as in Fig. 23a, located in the X-Y plane. It is desired to find the circulation about it, and this is done (in accordance with the definition of circulation) by multiplying the length of each side of the rectangle by the component of field strength parallel to that side. The lengths of the sides are (in the limiting case) dx and dy. Starting at the lower left corner in Fig. 23a, consider the bottom of the rectangle: the length is dx, in a positive direction, and the component of the field along the bottom of the rectangle is $E_{x \text{ at } y_1}$. The bottom of the rectangle therefore provides the first term of the following expression for circulation, the other three terms being obtained from the other three sides of the rectangle taken in order, counterclockwise:

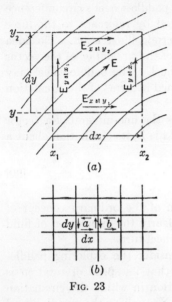

Fig. 23

$$\text{Circulation} = E_{x \text{ at } y_1} dx + E_{y \text{ at } x_2} dy - E_{x \text{ at } y_2} dx - E_{y \text{ at } x_1} dy \quad [92]$$

The third and fourth terms are negative because, in maintaining a counterclockwise direction about the rectangle, the top and left-hand side are traversed in a negative direction. The distance traveled along the top is $-dx$, and, along the left side, $-dy$.

Regrouping terms,

$$\text{Circulation} = (E_{y \text{ at } x_2} - E_{y \text{ at } x_1}) dy - (E_{x \text{ at } y_2} - E_{x \text{ at } y_1}) dx \quad [93]$$

The first parenthesis of the right-hand member is the amount by which E_y increases in the distance dx between x_1 and x_2; it is therefore equal to $\dfrac{\partial E_y}{\partial x} dx$. The quantity in the second parenthesis is similar and corresponds to the change of E_x in the distance y_1 to y_2. Hence equation 93 may be written

$$\text{Circulation} = \left(\frac{\partial E_y}{\partial x} dx\right) dy - \left(\frac{\partial E_x}{\partial y} dy\right) dx \quad [94]$$

Since, as seen above, circulation about an infinitesimal path is equal to $(\nabla \times \mathbf{E}) \cdot d\mathbf{a}$, **a** being the area within the path, we may write

$$(\nabla \times \mathbf{E}) \cdot d\mathbf{a} = \frac{\partial E_y}{\partial x} dx\, dy - \frac{\partial E_x}{\partial y} dy\, dx \qquad [95]$$

The differential area within the rectangular path is equal to $dx\, dy$, and both sides of the equation may be divided by this quantity. Since $d\mathbf{a}$ represents an area in the X–Y plane, equation 95 gives the component of curl normal to that plane. In the two-dimensional field this is the total curl, and

$$\nabla \times \mathbf{E} = \mathbf{k}\left(\frac{\partial E_y}{\partial x} - \frac{\partial E_x}{\partial y}\right) \qquad [96]$$

This is identical with equation 51. In a three-dimensional field this is one component only, the complete expression for curl being given by equation 47.

Stokes' Theorem. The discussion of the previous section concerns curl at a single point, or the region within an infinitesimal rectangle. It is now desired to determine the integral of curl over a surface of finite extent. This may be done by finding curl at every point of the surface and integrating. But there is a very helpful theorem, due to Stokes, that frequently saves a good deal of trouble.

Consider the small rectangle marked a in Fig. 23b, assumed to be in a vector field. There is circulation around this rectangle, as indicated by the arrows, corresponding to curl of the vector field within the area a. Now consider the adjoining rectangle marked b. There is circulation about this rectangle also, corresponding to curl in the area b. But, since the rectangles a and b have one side in common, this contributes a certain amount of circulation in one rectangle and an exactly equal but opposite amount in the other. Therefore, the sum of the curl in rectangle a plus the curl in rectangle b can be found by measuring the circulation around the outer perimeter of the larger rectangle made up of both a and b together, and giving no further attention to the equal and opposite components contributed by the common side.

Other rectangles may be added to these two, in any number. Always the circulation along common sides may be discarded, so that, no matter how large the final area or what its shape, the summation of curl at all points of a surface is equal to the circulation about the perimeter of the surface. Mathematically, the line integral of a vector, which defines circulation, may be equated to the surface integral of the curl of the vector, and the result is Stokes' theorem:

$$\oint \mathbf{E} \cdot d\mathbf{s} = \int (\nabla \times \mathbf{E}) \cdot d\mathbf{a} \qquad [97]$$

This theorem is not limited to a plane surface. Although the above discussion has been illustrated by reference to the plane surface of Fig. 23, the theorem applies and can be rigorously proved for a surface of any shape. The surface over which curl is integrated in equation 97 might, for example, be concave like a cup or a kettle. In the latter case the integration of curl all over the kettle would be equal to the circulation around the rim of the kettle.

But suppose the kettle has a lid, placed upon the kettle in the usual manner. The rim of the lid and the rim of the kettle coincide. Hence the integration of curl over the lid must be equal to the integration of curl over the kettle. That is to say, in more abstract terms, the integration of curl over a surface in a vector field, as indicated by the right-hand member of equation 97, is the same for all surfaces having a common perimeter and is quite independent of the shape of the surface.

Comparison of Theorems. Gauss's theorem and Stokes' theorem are very similar in their essential natures, for they relate large-scale phenomena to small-scale phenomena—the macroscopic to the microscopic.

If a vector field is examined minutely at a particular point, as with a microscope, it will be found to have a certain divergence at that point. This examination is performed, as a matter of fact, not with a microscope, but with a partial derivative, and equation 86 gives the divergence when the partial derivatives at the point are known.

But divergence has a large-scale result that can be detected without the aid of a microscope (or a partial derivative). This is flux: if flux issues from a volume, there is divergence within that volume, and Gauss's theorem gives the relation. One side of Gauss's theorem is in terms of the field passing through a surface; this is a surface of finite size and is the macroscopic quantity. The other side of the theorem is in terms of the divergence throughout a volume; this must be considered point by point and is the microscopic quantity.

Similarly, Stokes' theorem relates the macroscopic effect, circulation along a closed path, to the microscopic phenomenon, curl at every point of a surface bounded by that path.

These theorems are frequently useful when the microscopic nature of a field is known and the macroscopic nature is desired (that is, when one knows the derivatives and wishes to find the field) or vice versa. Such applications will be illustrated in the next chapter, in which our study of the electric field is continued.

Scalar Potential. The scalar field of elevation-above-sea-level is a potential field. It is a field of gravitational potential. Its value at any point is defined as the work required to move a body of unit mass to that point from sea level, or as the amount of potential energy gained by the body in being so moved.

SCALAR POTENTIAL

A level surface—or, more precisely, a surface of constant elevation above sea level—is an **equipotential surface,** for a body can be moved from one point to another of such a surface without any change of gravitational potential, and without any work being done. Equipotential surfaces near sea level are shown in Fig. 24a, and equipotential surfaces at greater distances from the earth in Fig. 24b.

The **gravitational field** is a field of force. It is therefore a vector field. Its value at each point is equal, by definition, to the gravitational force on a body of unit mass at that point. It is a field directed downward, toward the center of the earth.

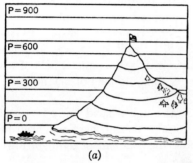 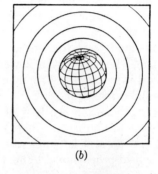

FIG. 24

Gravitational potential is defined in terms of work, and the gravitational field in terms of force, so there is a simple relation between them. The field is the negative of the *gradient* of the potential:

$$\mathbf{F} = -\nabla P \qquad [98]$$

Although written with reference to the gravitational field, this equation expresses a general relation between potential and force. The force field may be found from the potential field by differentiation. Every potential field has a gradient, and **F** can always be found from P. But it is not always possible, knowing a field of force, to find a corresponding potential field; a potential field does not always exist. Fortunately, there is a simple criterion by which we may know whether there is a potential field corresponding to a given field of force.

It was shown in Chapter II that, if a vector field is the gradient of some scalar field, the vector field has no curl. The converse of this theorem provides us with the desired criterion: *If a vector field is found to have no curl, that vector field is the gradient of some scalar field.* Symbolically,

$$\text{If } \nabla \times \mathbf{F} = 0 \text{ then } \mathbf{F} = -\nabla P \qquad [99]$$

This says that, if a vector field has no curl, some scalar field exists whose gradient is everywhere identical with the given vector field; it does not tell how to find that scalar field, but it does give assurance of its existence. When the scalar field P is found to exist, it is called a *potential* field.

The electrostatic field offers an example. As will be seen in the next chapter, the electrostatic field has no curl. Therefore we know from the above corollary that an electrostatic potential field exists. Indeed, it is well known that electrostatic potential exists, and that voltage is electrostatic potential difference.

The static magnetic field has no curl in regions that are not carrying current. This will be discussed in Chapter VI. Hence, if a wire is carrying current, there is a magnetic field around that wire which has no curl. A scalar field of magnetic potential can be found in the space around the wire. (If the wire is straight, the equipotential surfaces of this magnetic scalar-potential field will be radial planes.) But *within* the wire, where current is flowing, there will be a magnetic field of which the curl is not zero. Within the wire, then, no field of magnetic scalar potential can be found.[2]

As another example, flow of heat through a solid body is a vector field that has no curl. Therefore a scalar field of heat potential must exist. It does; it is temperature, and the gradient of temperature is a vector field proportional to heat flow.

In the gravitational field, one surface will contain all points at sea level (Potential = 0) and another all points 100 feet above sea level (Potential = 100) and so on. (See Fig. 24a and b.) By these equipotential surfaces, all space is divided into thin plates or shells called lamellas.

Only vector fields without curl have this characteristic of dividing space into lamellas (thin laminations) by means of equipotential surfaces, so, following the usage of James Clerk Maxwell, a vector field without curl is called a **lamellar** field. Sometimes a lamellar field is called **irrotational** because it has no curl.

Solenoidal Fields and Vector Potential. It was also seen in Chapter II that, if a vector field is the curl of another vector field, it has no divergence. Expressed in symbols:

$$\text{If } \mathbf{B} = \nabla \times \mathbf{A} \quad \text{then } \nabla \cdot \mathbf{B} = 0 \qquad [100]$$

This theorem also has its converse: *If a vector field has no divergence, that vector field is the curl of some other vector field.* Symbolically,

$$\text{If } \nabla \cdot \mathbf{B} = 0 \quad \text{then } \mathbf{B} = \nabla \times \mathbf{A} \qquad [101]$$

[2] The physical meaning of this is that, if a magnetic pole could be placed within the conductor (the conductor may be visualized as being mercury), the work done in moving the magnetic pole from one point to another would depend upon the path followed, and no value of potential can be assigned.

SOLENOIDAL FIELDS AND VECTOR POTENTIAL

In other words, if the vector field **B** has no divergence (as is true, for instance, in a field that represents the velocity of flow of an incompressible liquid), then some other vector field can be found, which we choose to call **A**, such that the curl of **A** is everywhere equal to **B**. At least, it is to be hoped that the field **A** can be found. The process of finding it is sometimes difficult and sometimes impossible, but, if **A** exists, whether or not its computation is feasible, it is given the name of **vector potential**.

The vector potential is somewhat analogous to the scalar potential of the previous section. Nevertheless, it must be emphasized that it is an entirely different quantity. If a vector field has neither curl nor divergence, it will have both a scalar potential and a vector potential, and they will be different quantities with little resemblance to each other.

A vector field without divergence is spoken of as **solenoidal or sourceless**. All the lines of flux are closed curves, having neither beginning nor end, a fact that follows necessarily from there being no divergence. Every solenoidal field has a vector potential. All magnetic fields are solenoidal, and therefore there is always a magnetic vector potential.

Example. Figure 16a, page 23, shows water circulating in a tub. It might equally well be interpreted as the magnetic field within a conductor of circular cross section. Assume a set of cylindrical coordinates; this is a three-dimensional set in which r is radial distance from an axis, z is distance parallel to the axis, and θ is angle as in polar coordinates (equation 69). The system of coordinates should be assumed with its axis coinciding with the axis of rotation of Fig. 16a; it is then the simplest and best adapted system to use for a problem of this character. (Formulas in terms of cylindrical coordinates are given in Table II.)

The field of Fig. 16a, whether considered to be velocity of water or magnetic field strength, is described by the equations

$$H_r = 0$$

$$H_\theta = ar \qquad [102]$$

$$H_z = 0$$

where a is a known constant. It is desired to determine the scalar and vector potentials of this field.

To determine whether a scalar potential exists, find the curl of the given field **H**. Since **H** is essentially two-dimensional, equation 82 is adequate (or the complete expression for curl in Table II may be used) and the result is

$$\nabla \times \mathbf{H} = \mathbf{k}\left(a\frac{\partial r}{\partial r} - 0 + \frac{ar}{r}\right) = \mathbf{k}2a \qquad [103]$$

Hence the curl is not zero, the field is not lamellar, and it is useless to try to find a scalar-potential field, for none exists.

To determine whether a vector potential exists, find the divergence of **H**. Using equation 81, each term of which is zero, it is evident that the field does not have divergence, and hence it should be possible to find a vector potential. Let us call the assumed vector-potential field **A**, and try to find what it is.

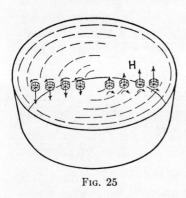

Fig. 25

The curl of **A** must be **H**, so, if little paddle-wheels—the "curl-meters" of Fig. 18—are placed with their axes along lines of **H**, as in Fig. 25, they must be turned at proper relative speed by the vector-potential field **A**. (As an alternative way of considering the problem, if all the curl-meters of Fig. 25 are driven at the proper speed, in proportion to the strength of **H** along their axes, they will act as pumps and, by churning the hypothetical liquid in which they are supposed to be, they will impart to it a velocity equivalent to the vector-potential field **A**.)

Now, by consideration of Fig. 25, it becomes clear that one possible solution for vector potential would be a vertical field that has zero intensity in the center of the tub and that increases in vertically downward intensity nearer the sides. This would spin all the curl-meters in the proper direction and suggests a vector-potential field parallel to the Z axis. The strength of this vector-potential field will vary with radius, but, since it will presumably be symmetrical, it will not vary with either θ or z. The manner in which the vector potential varies with radius is unknown and will just be expressed as a function of r, written symbolically $f(r)$. Thus

$$A_r = 0$$
$$A_\theta = 0 \qquad [104]$$
$$A_z = f(r)$$

This formulation for **A** is merely a guess. but if it is correct it will satisfy

SOLENOIDAL FIELDS AND VECTOR POTENTIAL

the requirement that $\nabla \times \mathbf{A} = \mathbf{H}$. For this to be true, each component of the curl of $\mathbf{A}$ must equal the corresponding component of $\mathbf{H}$. Writing out the components of $\nabla \times \mathbf{A}$ using Table II, and equating to components of $\mathbf{H}$ from equation 102, we find the following three equations that must be satisfied:

$$\frac{1}{r}\left(\frac{\partial A_z}{\partial \theta} - \frac{\partial(rA_\theta)}{\partial z}\right) = 0 \qquad [105]$$

$$\frac{\partial A_r}{\partial z} - \frac{\partial A_z}{\partial r} = ar \qquad [106]$$

$$\frac{1}{r}\left(\frac{\partial(rA_\theta)}{\partial r} - \frac{\partial A_r}{\partial \theta}\right) = 0 \qquad [107]$$

Equations 104 are tested by substitution into equations 105, 106, and 107. Substitution from equations 104 reduces equations 105 and 107 immediately to identities, and equation 106 becomes

$$-\frac{\partial A_z}{\partial r} = ar \qquad [108]$$

This is a differential equation, the solution of which is the desired function A_z. In this simple case, the solution is found by integration:

$$A_z = b - \frac{ar^2}{2} \qquad [109]$$

in which b is a constant of integration that may have any constant [3] value (but may not be a function of r or θ).

A_z from equation 109, and A_r and A_θ from equation 104, give the answer to our problem. They describe a field whose curl is $\mathbf{H}$. However, this is not the only field whose curl is $\mathbf{H}$. A field is not uniquely defined by specifying only its curl. It will be seen in a later chapter that divergence of $\mathbf{A}$ may also be specified, as well as curl. For the present it is enough to say that the vector-potential field $\mathbf{A}$ determined above, is of particular interest because it not only has the correct curl at every point, but it also has everywhere zero divergence.

[3] Indeed any irrotational function may be added to $\mathbf{A}$, for the curl of any irrotational function is zero.

PROBLEMS

1. Derive an expression for curl in polar coordinates, as in equation 82. Do not merely transform from an expression in rectangular coordinates, but start by summing circulation around an area such as that shown in the accompanying figure.

2. Figure 19a shows a vector field which, being without curl, must have a scalar potential. Assuming the potential at the origin of coordinates to be zero, find the potential at all other points. On a sketch of the field, draw lines (equipotential lines) connecting points of equal potential.

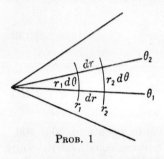

PROB. 1

3. Study Fig. 19b to determine whether equipotential lines can be drawn. Is there curl?

4. Does the vector field of Fig. 20 have a scalar-potential field? Can you determine it?

5. Gravitational potential is inversely proportional to distance from the center of the earth. What is the nature of the gravitational field, using equation 98?

6. Do equipotential surfaces exist within the canal of Fig. 17 (surfaces that define the scalar potential of the indicated velocity field)? Can you locate them?

7. A vector field $\mathbf{V}$ is defined by $V_x = 10$, $V_y = V_z = 0$. Find its scalar-potential field. Find its vector-potential field. Both vector and scalar potential are to be zero at the origin of coordinates.

8. Consider the possible existence of a vector field with zero divergence and zero curl at all points. Then impose the further restriction that the field strength must not anywhere be infinite. Then require also that the field strength be zero at an infinite distance in all directions.

9. Starting with an appropriate small section of space (as in Fig. 22a, but with shape adapted to the coordinate system), find the flux leaving such an elementary volume, and relate it to the expression (Table II, inside back cover) for divergence in spherical coordinates.

10. A vector field is defined by $E_r = B/r$, $E_\theta = 0$, $E_z = 0$ (cylindrical coordinates). Find a vector potential field $\mathbf{A}$ such that $\nabla \times \mathbf{A} = \mathbf{E}$ and $\nabla \cdot \mathbf{A} = 0$. Can $\mathbf{A}$ be zero at an infinite distance?

11. A vector field is defined by $E_r = B/r^2$, $E_\theta = 0$, $E_\phi = 0$ (spherical coordinates). Find a vector potential field $\mathbf{A}$ such that $\nabla \times \mathbf{A} = \mathbf{E}$, $\nabla \cdot \mathbf{A} = 0$, and $\mathbf{A}$ is zero at an infinite distance.

CHAPTER IV

The Electrostatic Field

In Chapter I, four experiments were described. In order to be useful, the results of these experiments were expressed in mathematical form as equations 2, 4, and 10. Equation 2 relates the force on a charged exploring particle to the electric field strength:

$$\mathbf{F} = Q\mathbf{E} \qquad [2]$$

Equation 4 states that the line integral of the electrostatic field about any closed path is zero:

$$\oint \mathbf{E} \cdot d\mathbf{s} = 0 \qquad [4]$$

Equation 10 states that the total flux leaving any closed surface is proportional to the excess positive charge within that surface:

$$\oint \mathbf{D} \cdot d\mathbf{a} = \oint \epsilon \mathbf{E} \cdot d\mathbf{a} = Q \qquad [10]$$

Equation 2 is found to be true *at all points*.

Equation 4 is found to be true along *every possible* closed path of integration.

Equation 10 is found to be true over *every possible* closed surface of integration.

Equations 2, 4, and 10 contain all necessary knowledge of electrostatics. No further experimentation is needed to develop the science of electrostatics in free space or isotropic material. Mathematical manipulation based on these equations will determine any electrostatic field when the charges that produce that field are known. The mathematics, however, is rarely easy. This chapter will be devoted to the fundamental principles of the mathematical solution, followed by a few examples.

Consider, first, equation 4. Applying Stokes' theorem (equation 97) gives:

$$\oint \mathbf{E} \cdot d\mathbf{s} = \int (\nabla \times \mathbf{E}) \cdot d\mathbf{a} = 0 \qquad [110]$$

The conclusion from this equation is that the electrostatic field has no curl. This conclusion would not result from any single measurement that gave a value of zero for the line integral of **E** around some one closed path, for such an experimental result would not preclude the possibility that curl might exist at various points in the field, even though the integral of curl over the surface of integration of equation 110 chanced to be zero. But it was specially emphasized that equation 4 applies to *all* closed paths, and therefore equation 110 applies to *all* surfaces, and the only way that equation 110 can apply to all surfaces of integration is for the curl of **E** to be zero everywhere. Hence

$$\nabla \times \mathbf{E} = 0 \qquad [111]$$

It follows from equation 111 that the electrostatic field is lamellar, and an electrostatic potential exists. If potential is called V, the electric field will (by equation 98) be the gradient of V (with a negative sign) so that

$$\mathbf{E} = -\nabla V \qquad [112]$$

Often in the solution of an electrostatic problem V can be found. It is then easy to find **E** by means of equation 112.

Another fundamental relation is expressed by equation 10. Applying Gauss's theorem to this equation gives

$$\oint \mathbf{D} \cdot d\mathbf{a} = \int (\nabla \cdot \mathbf{D}) \, dv = Q \qquad [113]$$

In any region in which there is no electric charge, so $Q = 0$, $\int (\nabla \cdot \mathbf{D}) \, dv = 0$, and hence the divergence of **D** is zero.

But, where charge is not zero, divergence is not zero. It is convenient to express divergence in terms of the density of electric charge (charge per unit volume) which may be called ρ. The charge within a closed surface is equal to the integral of the charge density through the contained volume. If the charge Q in equation 113 is expressed as $\int \rho \, dv$, the equation becomes

$$\int (\nabla \cdot \mathbf{D}) \, dv = \int \rho \, dv \qquad [114]$$

The two integrals of this equation are both volume integrals. Moreover, they are integrals throughout the same volume: through the volume, that is, contained within a specified closed surface. Finally, that closed surface is purely arbitrary; it may be any closed surface, of any size, shape, or location. Equation 114 can be true under this

CONDUCTORS

variety of conditions only if the integrand on one side of the equation equals the integrand on the other; if, that is,

$$\nabla \cdot \mathbf{D} = \rho \quad [115]$$

Within a homogeneous material, where ϵ does not change from point to point, this can be written

$$\nabla \cdot \mathbf{E} = \frac{\rho}{\epsilon} \quad [116]$$

This is an expression that relates electric field to charge density. It is also possible to relate electric potential to charge density by substituting equation 112 into 116, giving:

$$\nabla \cdot (\nabla V) = -\frac{\rho}{\epsilon} \quad [117]$$

or

$$\nabla^2 V = -\frac{\rho}{\epsilon} \quad [118]$$

When the meaning of the operator *nabla* is considered, it is evident that this is a second-order partial differential equation. It is of such importance that it is given a name: it is **Poisson's equation**. In the special case that applies to space containing no charge, it reduces to

$$\nabla^2 V = 0 \quad [119]$$

This is an even more famous differential equation, called **Laplace's equation**.

The study of electrostatics is essentially the solution of these equations. To sum up:

$$\mathbf{E} = -\nabla V \quad [112]$$

$$\nabla^2 V = -\frac{\rho}{\epsilon} \quad [118]$$

These relations completely define an electrostatic field, and their application to specific problems will now be considered.

Conductors. A conducting material is one in which electric charge can flow. Conducting materials contain electricity that is free to move when it is acted upon by the force of an electric field, and this is true even though they are "uncharged" in the sense of having no excess of either sign of charge.

Since charge can flow within a conductor, there can be no electrostatic field within the conducting material, for if there were it would exert force on the charge and move it from one point to another until the electric field was reduced to zero by the redistribution of the charge. Hence, within a conductor, in the electro*static* case:

$$\mathbf{E} = 0$$

from which: [120]

$$V \text{ is constant}$$

Since $\mathbf{E}$ is zero it follows that $\nabla \cdot \mathbf{E} = 0$, and hence that $\rho = 0$. Therefore, there can be no electric charge at any point within the material of a conductor. But $\mathbf{E}$ is not necessarily zero at the surface of a conducting body, and electric charge may be located on the surface. The conclusion is that all the charge on a conducting body will flow to the surface and remain there.

Certain general conclusions may be drawn about the electrostatic field in space just beyond the surface of a conductor. Since the entire conductor is at the same potential (equation 120), the surface of the conductor is an equipotential surface. The electric field is always normal to equipotential surfaces; electric field is the potential *gradient*, and gradient is always normal to the equipotential surfaces of the field from which it is derived, as considered in Chapter II. The electrostatic field at the surface of any conductor will therefore be normal to the surface of the conductor.

The previous paragraph gives the direction of the field at the conductor surface; something may also be said about the strength of the field. Since charge on the conducting body is distributed over the surface, it is convenient to speak of the density of charge in terms of charge per unit area. As ρ was used in equation 114 to represent charge per unit volume, σ will now be used to represent charge per unit area of surface, and total charge is found from

$$\int \sigma \, da = Q \qquad [121]$$

If the charge density per unit area of a conducting surface is σ, there must be σ flux lines extending normally outward from each unit area. This follows from the fact that one line emanates from each unit of positive charge. The electrostatic flux density in space just outside a charged conducting surface is therefore equal to the density of charge on the surface, and in direction it is normal to the surface. $\mathbf{D}$ is flux density, and its normal component at the surface (which may be written D_n) is

$$D_n = \epsilon E_n = \sigma \qquad [122]$$

A CHARGED SPHERE

So, if the distribution of charge on the surface of a conductor is known, the electric field just outside the conductor can be found. Similarly, if the field is known, the charge distribution can be found. Ordinarily, however, neither is known, and both must be found from the fact that the conductor surface is an equipotential surface with a known total charge, which makes the solution more complicated.

A Charged Sphere. Consider an isolated spherical conductor with a known charge Q upon it. Find the electric field about the sphere, in space that is filled with material of relative dielectric constant κ.

The problem may be solved by finding a potential field that satisfies Laplace's equation

$$\nabla^2 V = 0 \qquad [119]$$

and at the same time satisfies the boundary conditions that (1) the surface of the sphere is an equipotential surface, and (2) the total charge on the sphere is Q.

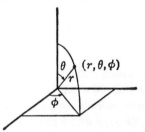

FIG. 26

The expansion of Laplace's equation in rectangular coordinates was given in equation 67, but, for use in a problem concerning a sphere, it will be much more convenient to expand in spherical coordinates. In using spherical coordinates, each point in space is located by a radial distance r and two angles θ and ϕ, as in Fig. 26. The derivation of the Laplacian in spherical coordinates will not be given here (it may be found in advanced calculus books); the result is included in Table II (inside back cover). It is

$$\nabla^2 V = \frac{\partial^2 V}{\partial r^2} + \frac{1}{r^2} \frac{\partial^2 V}{\partial \theta^2} + \frac{1}{r^2 \sin^2 \theta} \frac{\partial^2 V}{\partial \phi^2} + \frac{2}{r} \frac{\partial V}{\partial r} + \frac{\cot \theta}{r^2} \frac{\partial V}{\partial \theta} \qquad [123]$$

A solution of our problem, then, is an expression for potential that will make equation 123 equal zero and that at the same time will make the surface of the conducting sphere an equipotential surface.

Fortunately, the problem can be greatly simplified by consideration of symmetry. Since our charged sphere is isolated in space, whatever happens on or about the sphere must be independent of any direction except radial direction. No other direction can be defined. To distinguish any other direction, it would be necessary to have another object in space, for comparison.

So, if there is no distinction between different directions, the electric potential V about the sphere cannot be different in different directions.

It must be the same for all values of θ and ϕ (referring to Fig. 26 and assuming the center of the charged sphere at the origin of coordinates), varying only when r is changed.

Since V is a function of r only and does not vary with θ or ϕ, Laplace's equation reduces to

$$\nabla^2 V = \frac{\partial^2 V}{\partial r^2} + \frac{2}{r}\frac{\partial V}{\partial r} = 0 \quad [124]$$

Now this is an ordinary differential equation that is reducible to a linear equation with constant coefficients. The solution is

$$V = \frac{a}{r} + b \quad [125]$$

in which a and b are any arbitrary constants. To check the correctness of this solution, it may be substituted back into equation 123 which is thereby reduced to an identity.

Next, a and b must be evaluated from boundary conditions.[1] Assume that the potential at a very great distance from the sphere is unaffected by the charge on the sphere, so that when $r = \infty$, $V = 0$. Substituting these values into equation 125 gives $b = 0$, and the equation reduces to

$$V = \frac{a}{r} \quad [126]$$

The remaining constant a must be evaluated in terms of the charge on the sphere.

First, the radius of the sphere must be known; let it be r_0. Since the charge Q must be distributed symmetrically over the entire surface of the sphere, the charge per unit area σ is

$$\sigma = \frac{Q}{4\pi r_0^2} \quad [127]$$

---

[1] Solutions of differential equations, in general, contain terms that must be evaluated from boundary conditions. These are terms (constants or functions) that vanish when the solution is substituted into the differential equation, and hence they cannot be determined from the differential equation. But any physically useful solution must coincide with certain known boundary conditions, and this gives a means of evaluating the undetermined coefficients in a physical problem. For practical purposes, a solution of a differential equation that agrees with all initial and boundary conditions is a unique solution. Specifically, a unique solution of Laplace's equation is obtained if the Laplacian is everywhere zero and if the function meets the boundary conditions over a closed surface and vanishes at infinity. (See also footnote 8 on page 92.)

and from equation 122 it follows that the field strength at the surface of the sphere, radial in direction, will be

$$E = \frac{Q}{4\pi \epsilon r_0^2} \quad [128]$$

Now, since potential does not vary with θ or ϕ, the potential gradient is everywhere radial, and the magnitude of the electric field strength at any point is found [2] as

$$E = -\frac{\partial V}{\partial r} = \frac{a}{r^2} \quad [129]$$

This is true everywhere, so it is true at the surface of the sphere, and, equating 129 to 128 with the provision that $r = r_0$:

$$\frac{a}{r_0^2} = \frac{Q}{4\pi \epsilon r_0^2} \quad [130]$$

from which $a = Q/4\pi \epsilon$. Finally, then, at any point external to the charged sphere,

$$V = \frac{Q}{4\pi \epsilon r} = \frac{Q}{4\pi \kappa \epsilon_0 r}$$

$$E = \frac{Q}{4\pi \epsilon r^2} = \frac{Q}{4\pi \kappa \epsilon_0 r^2} \quad [131]$$

The physical interpretation of these quantities may well be repeated. E, the electric field, is a force; it is equal at any point to the force on an exploring particle with unit positive charge. V, the potential, represents work; the potential of a point in space is the work required to move to that point an exploring particle with unit positive charge, starting an infinite distance away.

The electric field is everywhere away from a positively charged body, because the force on a positive exploring particle will be repulsive; and potential increases as one approaches a positively charged body, because one must do work in moving an exploring particle against the force of the electric field.

Spherical Condenser. A charged conducting sphere A of radius a is concentrically located within a hollow conducting sphere B of inside

[2] See expression for gradient in spherical coordinates in Table II.

58   THE ELECTROSTATIC FIELD

radius b, as in Fig. 27. The electric field in the space between spheres A and B is identical with the field in the same region about an isolated sphere similar to A and with the same charge. It is radial in direction, and its strength is

$$E = \frac{Q}{4\pi \, \epsilon r^2} \qquad [132]$$

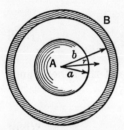

Fig. 27

The outer sphere B carries an electric charge equal and opposite to that on A, and the electric field terminates on the inner surface of B. The two spheres constitute a condenser; it has capacitance, and there is voltage between the spheres. Note that, with a given charge, the electric field strength E depends on κ, the dielectric constant of the material between the spheres, although the flux density D does not.

Voltage, which is essentially *potential difference* between two points (ordinarily between two metallic conductors), is defined as

$$V_{12} = \int_1^2 \mathbf{E} \cdot d\mathbf{s} \qquad [133]$$

That is, voltage from point 1 to point 2 is the line integral of the electric field along any path from point 1 to point 2. This is the amount by which point 1 is at a higher potential than point 2.

It is interesting to notice that potential difference or voltage between two points is the work that will be done by electric force on a unit electric charge that is allowed to move from one point to the other. Substituting equation 2 in equation 133,

$$V_{12} = \frac{1}{Q} \int_1^2 \mathbf{F} \cdot d\mathbf{s} \qquad [134]$$

and, since the integral of force times distance is work,

$$V_{12} = \frac{W_{12}}{Q} \qquad [135]$$

Voltage between spheres of the spherical condenser is found by locating a point 1 on the surface of A and a point 2 on the surface of B. The integration to determine voltage is simplest if 1 and 2 are on the same

POLARIZATION

radial line, for then the line of integration is parallel to the electric field, and all the distances are radial. In that case,

$$V_{AB} = \int_a^b E\, dr = \int_a^b \frac{Q}{4\pi\, \epsilon r^2}\, dr = \frac{Q(b-a)}{4\pi\, \epsilon ab} \qquad [136]$$

Capacitance of a condenser is, by definition, the charge divided by the voltage.

$$C = \frac{Q}{V} \qquad [137]$$

The capacitance of the spherical condenser is therefore

$$C = \frac{4\pi\, \epsilon ab}{b-a} \qquad [138]$$

a quantity that is determined entirely by the geometry and material of the condenser.

Occasionally one encounters references to the capacitance of an isolated sphere. This may be considered to be the limit approached by the capacitance of the spherical condenser as the outer sphere is allowed to become large without limit. Letting b in equation 138 approach infinity, we obtain, as the limiting value of capacitance,

$$C = 4\pi\, \epsilon a \qquad [139]$$

The capacitance of any condenser is proportional to ϵ, the dielectric constant. This fact suggests a means of measuring dielectric constants of materials, and such a means was actually used by Faraday and others in the classical determination of dielectric constants.

Polarization. It is convenient to speak of the relative dielectric constant of a material such as glass or oil or polystyrene and to use the appropriate value of κ in computation, but it is not clear without further consideration *why* the effect of one charged particle on another should depend on the nature of the material between them. Why is the electric field less in any solid or liquid dielectric material than in empty space?

It is supposed that all non-conducting material contains positive and negative charges bound together, perhaps by being part of the same atom. When material of this kind is in an electric field, the positive charges tend to move one way and the negative charges the other, but, since they are bound, they can move only as far as the elastic nature of the bond permits. Each atom is somewhat distorted, therefore, by the stress of the electric field and becomes positive on one side and negative on the other.

Figure 28 shows a block of dielectric material between a pair of charged metal plates, one with charge $+Q$ and the other with $-Q$. In the upper part of the diagram, the material is shown so much enlarged that the elementary particles can be seen; they are strained by the electric field so that each is negative on the left and positive on the right. The result is that the left-hand surface of the material is predominantly negative and the right-hand surface positive. Any cubic centimeter of the material, however, contains equal numbers of positive and negative elements and is neutral.

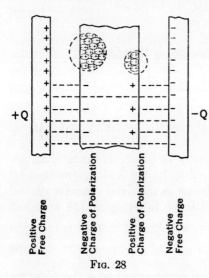

Fig. 28

In this way, without any flow of free charge through the material, but merely as a result of polarization, the material has acquired the equivalent of a surface charge. This explains why the electric field strength **E** is less in dielectric material than in free space; the surface charges of polarization partially shield the region within the material. Surface charge of polarization is not considered in any of the equations of this book; its effect is cared for by introducing the relative dielectric constant κ. It is possible, however, if desired, to develop a complete mathematical theory in terms of polarization, and this is done in comprehensive books on electrostatics.

Inverse Square Law. There is a radial electric field about a sphere with charge Q_1, the strength of which is given by equation 131 as

$$E_1 = \frac{Q_1}{4\pi \, \epsilon r^2} \qquad [140]$$

A second charged sphere, with charge Q_2, is moved into the electric field of the first sphere. It is required to find the force exerted on the second sphere by the electric field of the first.

By equation 2,

$$F = Q_2 E_1 \qquad [141]$$

and, if the distance from the center of one sphere to the center of the other is r, equation 140 gives

$$F = \frac{Q_1 Q_2}{4\pi \epsilon r^2} \qquad [142]$$

This is the well-known **Coulomb's law**. Historically, it was discovered in the latter part of the eighteenth century by direct experiment by Coulomb, who used his newly invented torsion balance. Much of the science of electrostatics was deduced from it. In the present discussion, however, Coulomb's law is itself deduced from the experiments described in Chapter I.

There is one important provision to be made in connection with equation 142. It is accurate only if the two charged spheres are so far apart that neither disturbs the distribution of charge on the surface of the other. For, if the charge on the first sphere were redistributed to any noticeable extent by the electrostatic attraction or repulsion of Q_2, the electric field about the sphere would no longer be strictly radial at all points. Perfect symmetry would not, then, exist. If the radius of each sphere is small compared with the spacing between spheres, no appreciable disturbance will take place and equation 142 will be accurate. For mathematical rigor, the limiting case is considered: the charges Q_1 and Q_2 are assumed concentrated at points, rather than being distributed on spheres. But this has the disadvantage of being physically impossible.

It may be emphasized that force is inversely proportional to ϵ and hence is less in any dielectric medium than in free space. Two charged bodies in oil, for example, would attract each other less strongly than in air.

Field within a Hollow Charged Sphere. Coulomb's law is quite difficult to substantiate experimentally with a high degree of accuracy, because the force that must be measured is small. The best verification of the law is based on the experimental evidence that the force on a charged exploring particle inside a hollow charged sphere is everywhere exactly zero. This measurement, made by Cavendish even before Coulomb's direct measurements of force and repeated with greater accuracy by Maxwell in the latter half of the nineteenth century, is an experimental proof of the inverse square law, for, if electric force from a point charge followed any other law, there would be a resultant force either toward or away from the walls of a hollow charged sphere. This can be proved by integrating to obtain the resultant force exerted on an exploring particle by charge distributed uniformly over the surface of the sphere.

In following the line of argument that has been developed in the preceding chapters, however, the conclusion that there is no electric field within a hollow conducting surface of any shape may be reached from Laplace's equation. Consider any closed conducting surface surrounding empty space. The surface, being conducting, must be equipotential. If any flux lines extend from the surface into the interior space, they must be normal to the surface. No flux lines can extend into the interior space and terminate there, for if they did there would be divergence in empty space, and this is inconsistent with Laplace's equation. No flux lines can start from an equipotential surface and return to that same surface; if they did there would be curl in the field, and this is not permitted by Laplace's equation. Therefore no flux lines can enter the interior space from the conducting surface, and there can be no electrostatic field in a cavity (that contains no charge) within a closed conducting surface.

The Potential Integral. An electric field can be found, if its potential field is known, by equation 112, and the distribution of electric charge can be determined from the electric field by equation 115. In homogeneous material these relations can be combined in Poisson's equation which relates charge distribution directly to the potential field:

$$\nabla^2 V = -\frac{\rho}{\epsilon} \qquad [118]$$

Since the Laplacian operation is a differentiation that is fairly easy to perform, we can readily find charge distribution in a known potential field.

The potential field, however, is not usually known as the starting point of a problem. More commonly the charge distribution is known and the potential field is to be found. Equation 118 then becomes a more or less complicated differential equation, a very simple example of which was the determination of the field about a charged sphere. If an explicit expression for potential in terms of charge could be found, not requiring the solution of a differential equation, it would be valuable. This can, indeed, be done and gives a result in the form of an integral.

By equation 131 the potential at a point that is r_1 meters distant from the center of a small sphere with electric charge Q_1 is

$$V = \frac{Q_1}{4\pi \, \epsilon r_1} \qquad [143]$$

If there is also a second charged sphere in the neighborhood, the potential at the point under consideration will be

$$V = \frac{Q_1}{4\pi \, \epsilon r_1} + \frac{Q_2}{4\pi \, \epsilon r_2} \qquad [144]$$

If there are many charged bodies, the potential at a point will be the sum of the potentials resulting from each:

$$V = \frac{1}{4\pi \, \epsilon} \Sigma \frac{Q}{r} \qquad [145]$$

Finally, if the charge is distributed in space with density (charge per unit volume) represented by ρ, the amount of charge in a differentially small volume dv will be $\rho\, dv$ and potential takes the form of an integral which sums the increments of potential resulting from each infinitesimal charge:

$$V = \frac{1}{4\pi\,\epsilon} \int \frac{\rho\, dv}{r} \qquad [146]$$

This is the desired expression for potential in terms of charge. The meaning is clear from the derivation: the numerator of the fraction is

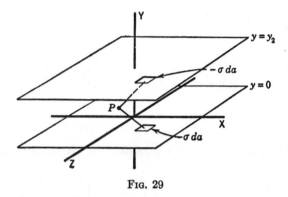

Fig. 29

an infinitesimal bit of charge, r is the distance from that bit of charge to the point at which potential is being determined, and the integration gives the summed effect of all charges present. The limits of the integration are such as to include all the region containing charge; if charge were distributed through a spherical region, for example, the integration would be throughout the sphere. If charge were distributed on the surface of a sphere, however, the differential element of charge would be $\sigma\, da$ rather than $\rho\, dv$ in equation 146, and the integration would only need be over the surface of the sphere.

The distance r depends upon the coordinates of the point at which V is being determined, and also the coordinates of the point at which each bit of charge is located. The formulation of an expression for this distance is unfortunately so complicated that the integration is impractical except for the simplest electric fields. However, the general method is of great importance in determining the radiation from antennas. In preparation for that use, a simple example will be given here.

Example. Consider the electric field between two uniformly charged parallel planes, as in Fig. 29. The planes are of infinite extent; the lower plane has a charge of σ per unit area, and the upper of $-\sigma$ per unit area. Coordinates are chosen with the X and Z axes lying in the lower plane. Point P is any point between the planes, and its coordinates are x_1, y_1, and z_1. In an infinitesimal area da located at $(x,0,z)$ in the lower plane, there is electric charge $\sigma\, da = \sigma\, dx\, dz$, and, directly above it in the upper plane at a point (x,y_2,z), there is an opposite charge $-\sigma\, dx\, dz$. The distance from the electric charge in the lower plane to the point P is $\sqrt{(x-x_1)^2 + y_1^2 + (z-z_1)^2}$, and from the electric charge in the upper plane to P is $\sqrt{(x-x_1)^2 + (y_2-y_1)^2 + (z-z_1)^2}$. The integral can now be written:

$$V = \frac{1}{4\pi\epsilon} \int \frac{\sigma\, da}{r}$$

$$= \frac{\sigma}{4\pi\epsilon} \int_{-\infty}^{\infty} \int_{-\infty}^{\infty} \left[\frac{1}{\sqrt{(x-x_1)^2 + y_1^2 + (z-z_1)^2}} - \frac{1}{\sqrt{(x-x_1)^2 + (y_2-y_1)^2 + (z-z_1)^2}} \right] dx\, dz \quad [147]$$

The process of integration is moderately involved, but the result is simple:

$$V = -\frac{\sigma}{\epsilon}\left(y_1 - \frac{y_2}{2}\right) \quad [148]$$

This same answer could have been obtained, and much more easily, by solution of Laplace's equation (see Problem 2, page 67).

Electrostatic Energy. A certain amount of energy must be expended to produce an electrostatic field. An exactly equal amount of energy is released when the electrostatic field ceases to exist. While the electrostatic field exists, this energy is stored in the field (or, at least, this is the customary assumption). It is even possible to determine the distribution of this energy in the field.

Consider two conducting bodies of any shape, such as A and B of Fig. 30. At first they are uncharged, but electric charge is gradually removed from B and added to A. Because of electrostatic force, a certain amount of work must be done in transferring charge from one body to the other. Since energy is the line integral of force, and potential is the line integral of electric field, it follows from equation 2 that the incremental energy required to transfer a small charge dQ through a potential difference V between the bodies is

$$\text{Incremental energy} = V\, dQ \quad [149]$$

ELECTROSTATIC ENERGY

But V can be expressed in terms of the capacitance of the bodies and of the charge that has already been placed on A before the small charge dQ is transferred, giving

$$\text{Incremental energy} = \frac{Q}{C} dQ \qquad [150]$$

The total energy expended in placing a total charge Q upon the bodies

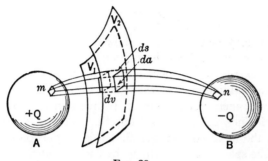

Fig. 30

is the summation of the small amounts of energy required by the small increments of charge, so the total

$$\text{Energy} = \int_0^Q \frac{Q}{C} dQ = \frac{Q^2}{2C} \qquad [151]$$

Since C is, by definition, Q/V, this may be written

$$\text{Energy} = \tfrac{1}{2} QV \qquad [152]$$

The total energy of the system is therefore one-half the product of *the charge on one of the bodies* and *the potential difference between the bodies*. The total energy can be subdivided to determine its distribution in space.

First, the energy corresponding to a given amount of charge is proportional, as in equation 152, to the potential difference through which the charge is moved. Let V_1 and V_2 be the potentials of equipotential surfaces enclosing the body A. A part of each of these surfaces is shown in Fig. 30. These could be actual thin metallic surfaces without altering the electric field. The charge $+Q$ might then be moved from the body A to the surface with potential V_1 (which completely encloses A) and the electric field between A and V_1 would then cease to exist. The electric field between the surface at V_1 and the body B, however, would be unchanged. The surface at V_1 would still be, as before, an equipotential surface; the same amount of flux would issue from it and would have the same distribution. Now let the charge be moved to the other equipotential surface V_2; the field from V_2 to B will remain unchanged, but the field from V_1 to V_2 will be eliminated. At the same time the energy of the system will be reduced from $\frac{1}{2}QV_1$ to $\frac{1}{2}QV_2$ (V_1 and V_2 being the potentials of the respective surfaces relative to B). Hence it is reasonable to say that the energy stored in the space between those surfaces was

$$\frac{1}{2}Q(V_1 - V_2) \qquad [153]$$

This discussion may be extended to show a division of energy in shells between equipotential surfaces throughout the entire electrostatic field.

Second, the energy of the field may be distributed among the flux lines. If each one-hundredth of the charge on A is responsible for one-hundredth of the energy of the system (which is reasonable from equation 152) and since there issues from that part of the charge one-hundredth of the flux, we may say that this proportion of the energy is located in the space through which the corresponding flux lines pass. In Fig. 30 a tubular section of the space between the charged bodies is indicated as containing flux passing from the small area m on A to n on B. The energy in the tube of space from m to n is proportional to the number of flux lines traversing this tube.

If the tube from m to n is so slim that its cross-section area where it intersects the equipotential surface V_2 is da, and the electric flux density at this cross section is $\mathbf{D}$, the amount of flux in the tube is $\mathbf{D} \cdot d\mathbf{a}$. The total electrostatic flux is Q. The energy in the tube is $\mathbf{D} \cdot d\mathbf{a}/Q$ times the total energy, or, from equation 152,

$$\frac{1}{2}(\mathbf{D} \cdot d\mathbf{a})V_{AB} \qquad [154]$$

If this tube of differential cross-section area is now cut by equipotential surfaces V_1 and V_2, the energy in the tube between those surfaces is, by equations 153 and 154,

$$\frac{1}{2}(\mathbf{D} \cdot d\mathbf{a})(V_1 - V_2) \qquad [155]$$

Finally, if the spacing between equipotential surfaces is reduced to the differential distance $d\mathbf{s}$ through which the potential drop $V_1 - V_2 = \mathbf{E} \cdot d\mathbf{s}$, the energy in the differential volume of space dv is

$$\frac{1}{2}(\mathbf{D} \cdot d\mathbf{a})(\mathbf{E} \cdot d\mathbf{s}) = \frac{1}{2}(\mathbf{D} \cdot \mathbf{E})\,dv = \frac{\epsilon}{2}E^2\,dv \qquad [156]$$

This may be interpreted to mean that the energy density (energy per unit volume) at any point is $\frac{1}{2}\mathbf{D} \cdot \mathbf{E}$ or $\frac{1}{2}\epsilon E^2$. Also it indicates that energy in any given region of space is obtained by integrating through that region:

$$\text{Energy} = \tfrac{1}{2} \int \mathbf{D} \cdot \mathbf{E}\, dv \qquad [157]$$

Energy in the entire electric field is found by integrating through infinite space.

That this infinite integral does, indeed, equal the total energy given by equation 152 can be proved directly. But it must not be overlooked that there is no definite proof that the energy *is* stored in space as suggested by equation 156—we can only say that it behaves as it would *if it were* so stored. The distinction is fundamental and is related to the great controversy between "field theory" and "theory of action at a distance" that raged during the nineteenth century. More will be said of this controversy in a later chapter.

PROBLEMS

1. A charge Q is placed upon an isolated metal sphere of radius r_0. Find electric field strength and potential at all points within the sphere.

2. Two parallel conducting plane surfaces form a condenser. Solve Laplace's equation for the electric field between the surfaces when they have charge ω and $-\omega$ per unit area. Find capacitance per unit area. (Consider a region that is so far from the edges of the surfaces that the electric charge is uniformly distributed.)

3. Find the electric field between two charged plane conducting surfaces set at an angle α, but not quite touching. Find the distribution of charge on the surfaces. Consider, as in Problem 2, a region distant from the edges, so that the result is not influenced by edge effects. Voltage between the plates is V.

PROB. 3

4. Find the electric field about an isolated cylindrical conductor of radius r_0 and unlimited length with a charge of q units per unit length. Find the curl of this electric field.

5. Find the potential difference between the conductor of Problem 4 and any point in space.

6. Find a law (similar to the inverse square law for charged spheres) to be used for force per unit length between charged parallel cylinders of unlimited length.

7. Find the energy required to charge unit area of the condenser of Problem 2, using equation 152. Find the energy in the electrostatic field of that condenser, per unit volume, using equation 156.

8. Two parallel metal plates are close together but insulated. They are charged, one being positive and the other negative. They are connected by flexible wires to an electroscope. Without permitting any change in the charge on either plate, the plates are separated. Explain why the electroscope indicates a greatly increased voltage between plates as the plates are drawn apart. If the initial separation is 0.1 millimeter and the initial voltage difference 100 volts, find the voltage between

them when they are separated 10 centimeters. (Note: This explains the high voltages of lightning.)

9. Find the capacitance per square centimeter of a pair of large parallel plates 1 centimeter apart in air. Convert the result to micromicrofarads (10^{-12} farad). Remember this result, at least approximately, for it is often useful in estimating capacitances.

10. Find the capacitance of a spherical condenser as in Fig. 27 if the space surrounding the inner sphere A is filled with paraffin out to a radius equal to $\frac{1}{2}(a + b)$. Find the electric field in both paraffin and empty space that results from a charge Q on the condenser.

11. Referring to Problem 8: if the initial separation is 0.1 millimeter of *paraffined paper* with a dielectric constant of 2.3, find the voltage when the plates are separated by 10 centimeters of air.

12. Write in detail the integral for potential (equation 146) about an isolated charged sphere, with radius r_0 and charge Q. Do not integrate.

13. From equation 148, find E between the condenser plates in the example of Fig. 29. Find the voltage difference between the plates.

CHAPTER V

Electric Current

In electrostatic problems, materials are divided into two classes: conductors and non-conductors. Electrostatics is not concerned with the relative ability of different conducting materials to permit charge to flow, for if there is any motion of charge the condition is not electrostatic.

If an electric field exists in a conducting material, electric charge in the material will be driven to flow in the general direction of the electric field. Such flow of charge is called an **electric current**; the electric current through any given surface is equal, by definition, to the rate of flow of electric charge past that surface. That is,

$$I = \frac{dQ}{dt} \qquad [158]$$

in which Q is the total charge that has passed through the surface.

An experiment is now necessary to determine the current-carrying characteristics of various conducting materials. This experiment, which will be called Experiment V, is actually the work of Georg Ohm, and the result is **Ohm's law**.

EXPERIMENT V. Current flowing in a metallic conductor is measured, and the voltage difference is determined between the ends of a section of the conductor (each end of the section is an equipotential surface). The voltage is maintained constant while the experiment is performed. This is done for an unlimited variety of sizes and shapes of conductors, and for many different metallic materials. It is found that the current and voltage are always related by

$$I = \frac{1}{R} V \qquad [159]$$

where R is a constant determined by the composition and geometry of the conductor.[1] It is found, moreover, that R is directly proportional

---

[1] Temperature and other physical conditions affect R to some extent. The essential point is that a given sample of material has a certain resistance which is a characteristic of the material and is not dependent upon the amount of current flowing or the voltage applied.

to the length of a conductor of constant cross section, and inversely proportional to the cross section of a conductor of constant length, so, with γ as a coefficient that is characteristic of the conductor material,

$$\frac{1}{R} = \gamma \frac{\text{area}}{\text{length}} \qquad [160]$$

The coefficient γ is called the **conductivity** of the material.

Voltage was defined by equation 133 as the integral of electric field strength:

$$V = \int \mathbf{E} \cdot d\mathbf{s} \qquad [161]$$

It is now desirable to introduce a new term, **current density ι**, which is so defined that the integral of it over a surface gives the current through that surface:

$$I = \int \iota \cdot d\mathbf{a} \qquad [162]$$

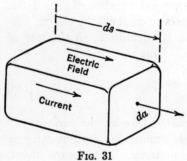

Fig. 31

Since Ohm's law applies to all shapes and sizes of conductors, it applies to a section of differentially small size. Consider the section shown in Fig. 31. The current flowing through this section is $\iota \cdot d\mathbf{a}$ (from equation 162) and the voltage from end to end is $\mathbf{E} \cdot d\mathbf{s}$ (from equation 161) and

$$\frac{1}{R} = \gamma \frac{da}{ds} \qquad [163]$$

from equation 160. Combining these in Ohm's law, equation 159,

$$\iota \cdot d\mathbf{a} = \gamma \frac{da}{ds} \mathbf{E} \cdot d\mathbf{s} \qquad [164]$$

Since $d\mathbf{s}$ and $d\mathbf{a}$, considered as vectors, have the same direction as $\mathbf{E}$ and ι, current density and electric field strength are related simply as

$$\iota = \gamma \mathbf{E} \qquad [165]$$

This resulting equation is a *microscopic* Ohm's law. It says that the current density at any point in a conductor is proportional to the electric field strength at that point, and in the same direction. This is true for metallic conductors, to which the above discussion has been limited. It is true for electrolytic conductors, also, and for many other conducting materials. (But it is not true for certain crystalline substances, such as carborundum, in which current is not proportional to voltage, nor to anisotropic material in which conductivity is not the same in different directions, nor to the passage of electricity through gases; Ohm's law does not apply to these cases without special interpretation. Cases in which equation 165 does not apply will not be considered further.)

When the electric field is not varying (as was specified in Experiment V) there can be no change of accumulated charge, for, if there were a change of charge, there would be a change of electric field also. It follows that whatever electrostatic charge there may be on the surface of the conductor will remain there, unchanged, while current flows through the cross section of the conductor. To repeat, since there is no change of electric field, there is no change of charge density at any point. Since there is no change of charge density at any point, the lines of current flow do not terminate. Since the lines of current flow do not terminate, the vector field of current density has no divergence. That is,

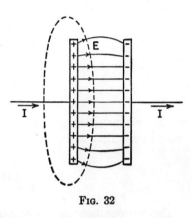

Fig. 32

$$\nabla \cdot \iota = 0 \qquad [166]$$

or, from equation 165, within the conductor,

$$\nabla \cdot \gamma \mathbf{E} = \gamma \nabla \cdot \mathbf{E} = 0 \qquad [167]$$

when the flow of current is steady and the electric field is unchanging.

Kirchhoff's first law may be stated: The algebraic sum of the currents flowing toward a junction in an electrical network is zero. This is clearly a special case of the more general relation of equation 166 and may be considered to be derived from it.

If, as a result of current flowing, charge is accumulating in some part of the circuit, equations 166 and 167 do not apply. Consider, for instance, the parallel-plate condenser of Fig. 32. Current is flowing into the left-hand plate, and from the right-hand plate. No current, however, is flowing in the space between the plates, but the electric field **E** in that region is constantly increasing as charge is deposited on one plate of the condenser and removed from the other.

Now imagine a closed surface about one of the plates, as indicated by the dash line. Charge is entering this closed surface, because of the current I, but no charge is leaving it. Consequently the amount of charge within the surface is increasing, and therefore the amount of electric flux passing out through the surface is increasing.

The rate of increase of charge within the surface is I, for, by equation 158, $I = dQ/dt$. Since one line of flux emanates from each unit of charge, the rate of increase of flux passing through the surface is

$$\frac{d}{dt}(\text{flux}) = \frac{d}{dt}Q = I \qquad [168]$$

However,

$$\text{Flux} = \int \mathbf{D} \cdot d\mathbf{a} \qquad [169]$$

so the current entering any closed surface is related to the flux passing out through that surface by

$$I = \frac{d}{dt}\oint \mathbf{D} \cdot d\mathbf{a} = \oint \frac{d\mathbf{D}}{dt} \cdot d\mathbf{a} \qquad [170]$$

The current flowing out through a closed surface is found by integrating current density over that surface:

$$\oint \mathbf{\iota} \cdot d\mathbf{a} \qquad [171]$$

and, since equation 170 is for current *entering*, we may equate

$$-\oint \mathbf{\iota} \cdot d\mathbf{a} = \oint \frac{d\mathbf{D}}{dt} \cdot d\mathbf{a} \qquad [172]$$

The two integrations of equation 172 are performed over the same closed surface, so

$$\oint \left(\mathbf{\iota} + \frac{d\mathbf{D}}{dt}\right) \cdot d\mathbf{a} = 0 \qquad [173]$$

whence, by Gauss's theorem,

$$\int \nabla \cdot \left(\mathbf{\iota} + \frac{d\mathbf{D}}{dt}\right) dv = 0 \qquad [174]$$

and, since this is true for the space contained within any closed surface, however large or small, it follows that everywhere the quantity in parentheses has zero divergence:

$$\nabla \cdot \left(\iota + \frac{d\mathbf{D}}{dt} \right) = 0 \qquad [175]$$

Two obvious substitutions in equation 175 then give

$$\nabla \cdot \left(\gamma \mathbf{E} + \epsilon \frac{d\mathbf{E}}{dt} \right) = 0 \qquad [176]$$

Comparison of equation 175 with 166 is very enlightening. Equation 166 tells us that current has no divergence in steady flow, that is, if the electric field strength is unchanging. If the electric field is changing, however, current does not flow without divergence. But, as we are informed by equation 175, if another term involving the rate of change of the electric field is added to current density at every point, the result is a quantity that has zero divergence under all circumstances.

We are tempted to look upon this additional quantity, this $d\mathbf{D}/dt$, as something similar to current—perhaps even as a kind of current itself. It is not a **conduction current,** which is the name given to $\gamma \mathbf{E}$, so we will call it a **displacement current.** The total current, the sum of conduction current and displacement current, according to this terminology, is made up of two parts:

$$\iota_t = \iota_c + \iota_d = \gamma \mathbf{E} + \frac{d\mathbf{D}}{dt} \qquad [177]$$

The divergence of this *total* current density is always zero, and ι_t is therefore *solenoidal*.

Electromotive Force. The electric field **E**, as it has been considered so far in the discussion, results from the presence of free electric charge. The flux lines of the field begin and end on electric charge. But any source of electromotive force can also contribute to the electric field: a familiar and important example that will receive a good deal of attention in later chapters is the induced electric field that appears in the neighborhood of a changing magnetic field. If the component of electric field that is due to electric charge is called $\mathbf{E}_s$ and the component resulting from electromotive action is called $\mathbf{E}_m$, then we may continue to use **E** for the total electric field strength at any point, and

$$\mathbf{E} = \mathbf{E}_s + \mathbf{E}_m \qquad [178]$$

Voltage, as that term is commonly used, is the line integral of $\mathbf{E}_s$. Voltage was defined with this meaning in equation 133, which should now be written more explicitly as

$$V = \int \mathbf{E}_s \cdot d\mathbf{s} \qquad [179]$$

Electromotive force is a similar line integral of $\mathbf{E}_m$:

$$\text{Electromotive force} = \int \mathbf{E}_m \cdot d\mathbf{s} \qquad [180]$$

As mentioned above, electromotive force may result from magnetic action, as in a generator or a transformer. It may also come from

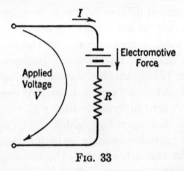

Fig. 33

chemical action in a battery, or from heat in a thermocouple, or from various other physical processes. In general, an electromotive force appears when energy of some other kind, such as chemical energy, or heat energy, or mechanical energy, is changed into electric energy.

Ohm's law is written in equation 159 for a section of circuit in which there is no electromotive force, and current flows as a result of applied voltage only:

$$IR = V = \int \mathbf{E}_s \cdot d\mathbf{s} \qquad [181]$$

But, in a part of a circuit where there is electromotive force, the electromotive force will contribute to the flow of current, and

$$IR = V + \text{Electromotive force} \qquad [182]$$

Thus, in the circuit of Fig. 33, the applied voltage is V, the resistance is R, and the electromotive force of a battery assists the flow of current. Current is therefore

$$I = \frac{V + \text{Electromotive force}}{R} \qquad [183]$$

as in equation 182. Equation 182 may also be written

$$IR = \int \mathbf{E}_s \cdot d\mathbf{s} + \int \mathbf{E}_m \cdot d\mathbf{s} = \int \mathbf{E} \cdot d\mathbf{s} \qquad [184]$$

Voltage in Fig. 33 is mathematically the line integral of $\mathbf{E}_s$ between the upper and lower terminals, following any desired path of integration. These terminals and the wires connected to them are, of course, charged bodies. If the path of integration is so chosen that it does not pass through any region in which any electromotive force exists, so that $\mathbf{E}_m$ is zero, the voltage is equally well expressed as the line integral of $\mathbf{E}$, as may be seen from equation 184, for, in the absence of electromotive force, $\mathbf{E} = \mathbf{E}_s$.

A particularly interesting special case appears in a closed circuit containing electromotive force. If the section of circuit under consideration in equations 182 and 184 is expanded until it becomes the whole circuit, and if the two terminal points approach each other until they coincide, the paths of integration of the line integrals become closed paths. But the line integral of $\mathbf{E}_s$ about any closed path is zero (equation 4) and equation 184 thus reduces, for a closed circuit, to

$$IR = \text{Electromotive force} = \oint \mathbf{E} \cdot d\mathbf{s} \qquad [185]$$

This relation will be of use in the next chapter, in exploring the magnetic field.

It may be noted in passing that the term that appears in ordinary alternating-current circuit computations as "inductive reactance voltage drop" is, in equations 182 and 184, part of the electromotive force term, for it is fundamentally a "counter electromotive force" produced by changing magnetic field. If there is a condenser in a circuit, its voltage is part of the $\int \mathbf{E}_s \cdot d\mathbf{s}$ term.

PROBLEMS

1. Find, from tables or otherwise, the conductivity of silver, gold, copper, iron, and aluminum. Give values in mhos per meter. What is the mks unit of resistivity?

2. Derive Kirchhoff's first law, as stated on page 71, from equation 166. Kirchhoff's second law says, in effect, that, in the absence of electromotive forces, the sum

of the voltages around any circuit of a network is zero; derive this law, also, as a special case of an equation in Chapter V.

3. A flat metal plate of uniform thickness is bounded by two quarter-circles and two radial lines, as shown in the figure. A constant direct voltage is maintained at

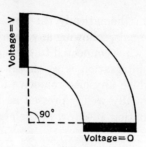

Prob. 3

the edges bounded by radial lines. Find the distribution of current density in the plate.

4. Thickness of the plate of Problem 3 is $\frac{1}{16}$ inch, the inner edge has 1-inch and the outer edge 2-inch radius of curvature. Total current is 100 amperes. Find the maximum current density in the plate.

5. The material of the plate of Problems 3 and 4 is aluminum. Find the voltage V.

6. From the terminals of a 10-kilovolt-ampere alternating-current generator, two parallel wires extend 500 yards to a load consisting of an electric heater, a static condenser, a magnetic relay coil, and an induction motor all connected in parallel. Using the definitions of the section on electromotive force, where does E_m exist? Where is there E_s? Where can voltage be measured accurately with an ordinary voltmeter? Indicate in a diagram the flux lines of E_s. Can flux lines of E_m be drawn? Explain.

CHAPTER VI

The Magnetic Field

Magnetic Force. When a conductor is carrying current, there may be a mechanical force exerted upon it. This is quite distinct from electrostatic force and from all non-electrical forces, for it disappears when current ceases to flow. This force is observed when the conductor is in the neighborhood of another conductor that is also carrying current or when it is in the neighborhood of a magnet. It is therefore called magnetic force.

EXPERIMENT VI is performed to study magnetic force. The apparatus is indicated in Fig. 34. A short, straight section of conducting wire is mounted in such a way that force exerted upon it can be measured while current is flowing through it from end to end. Since the short section of wire must be free to move, in order to measure force, some kind of flexible connection is used to carry current to it. An arrangement of pools of mercury might well be employed.[1]

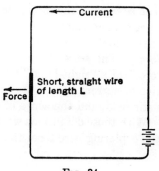

FIG. 34

The experiment shows that magnetic force on the exploring wire is always normal to the wire. The amount of the force is proportional to the amount of current flowing through the wire. The force is also proportional to the length of the exploring section of wire. All these factors are easily understood, for they depend upon the exploring wire and the current in the exploring wire. But the amount and direction of the magnetic force are also dependent upon the location and the orientation of the exploring wire in space, particularly with reference to magnets and to other circuits carrying electric current. This suggests that there is some condition in space (especially in the space about magnets and

---

[1] Historically, various forms of this apparatus were used by Ampère in 1821 in establishing the law of force between conductors that is known as "Ampère's law."

electric currents) which produces the magnetic force. It suggests that we would do well to consider the possible existence of a **magnetic field**.

The experimental evidence tells us that at any point in space it is possible to orient the exploring wire in such a way that there is no magnetic force upon it. If the exploring wire is held at the same point, but turned to a new orientation, there is then a magnetic force; and the amount of the force is proportional to the sine of the angle between the direction of the exploring wire and its direction when the force is zero. The magnetic force reaches a maximum when the wire is perpendicular to its null direction. The direction of the magnetic force, in addition to being normal to the wire, is also normal to the null direction.

From these experimental results it is seen that the idea of a magnetic field is quite reasonable. It must be a vector field with direction as well as magnitude. There is only one uniquely defined direction: the direction of the exploring wire when the magnetic force upon it is zero. This is taken, by definition, to be the *direction* of the magnetic field. The *strength* of the magnetic field is found from the maximum magnetic force that appears when the exploring wire is normal to the null position; the magnetic field strength is defined as proportional to this maximum force. The *sense* of the field is also defined in terms of this maximum force, for a "right-hand" relation is assumed between the positive direction of current flow in the exploring wire, the positive direction of magnetic field, and the sense of the resultant force.

With these definitions we are able to compute the magnetic force upon the exploring wire from the following equation:

$$\mathbf{F} = I\mathbf{L} \times \mathbf{B} \qquad [186]$$

The force is represented by $\mathbf{F}$, $\mathbf{L}$ is the length and direction of exploring wire on which the force is exerted, and I is the current that it carries. $\mathbf{B}$ is called the **magnetic induction** (as will be seen later, it is also the **magnetic flux density**). With $\mathbf{F}$ in newtons, $\mathbf{L}$ in meters, and I in amperes, $\mathbf{B}$ is measured in webers per square meter; no factor of proportionality is required in the equation.[2] It will be noticed that the current is written as a scalar quantity, whereas $\mathbf{L}$, the length of the wire that carries the current, is a vector, and its direction is obviously the

[2] Equation 186 is correct also for the "electromagnetic" cgs system of units: $\mathbf{F}$ in dynes, $\mathbf{L}$ in centimeters, I in abamperes, $\mathbf{B}$ in gausses. The Gaussian system, which uses the same cgs units except that I is in statamperes, requires a factor of proportionality (between abamperes and statamperes) usually designated c and equal to 2.998×10^{10}; with Gaussian units

$$\mathbf{F} = \frac{I\mathbf{L} \times \mathbf{B}}{c}$$

MAGNETIC FLUX

direction of the wire. The positive sense of **L** is arbitrarily selected, and when current flows in the positive direction it is positive current.

Equation 186 takes into account all the facts discovered by Experiment VI. The reader may well review the experimental results as described above and see how they are incorporated in the equation. The similarity of equation 2 for electrostatics and equation 186 for magnetostatics is interesting; so, also, are their differences.

$\vec{F} = Q\vec{E}$

Magnetic Flux. Lines of **magnetic flux** may be conceived just as were lines of electrostatic flux, and the definition is similar.

$$\text{Magnetic flux} = \Phi = \int \mathbf{B} \cdot d\mathbf{a} \qquad [187]$$

The quantity **B** is magnetic flux density, for, when multiplied by an area (or integrated over an area, as in equation 187), the product is flux. The mks unit of flux is the weber (1 weber is 10^8 maxwells) and hence the mks unit of flux density is, as in equation 186, the weber per square meter.[3]

EXPERIMENT VII. Another experiment will now be performed to study the relation between the magnetic and the electric fields. The discovery that an electric field could be produced magnetically was made by Michael Faraday in England in 1831, and it is often considered to be the most significant of his many valuable experiments. It was made independently,

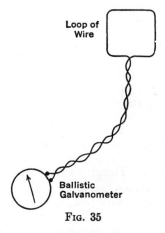

FIG. 35

but a few months later, by Joseph Henry in the United States.

The apparatus for Experiment VII is a loop of wire connected to a ballistic galvanometer. The arrangement of the apparatus may be as in Fig. 35 in which a twisted pair of wires is indicated between the loop and the galvanometer. The reading of a ballistic galvanometer is a measure of the electric charge that passes through it, and it is found that, if the loop is in a magnetic field, the galvanometer indicates that charge flows whenever the magnetic field strength is increased or decreased. We measure the magnetic flux density at the loop by the methods of Experiment VI and find that the reading of the ballistic galvanometer is proportional to the increase or decrease of flux passing through the loop of our apparatus. It is also determined that deflection of the gal-

---

[3] The weber per square meter, being 10 kilogausses, is a unit of convenient size.

vanometer is inversely proportional to the total resistance of the apparatus including loop, leads, and galvanometer. Since the galvanometer measures electric charge, we write

$$Q = -\frac{\Phi}{R} \qquad [188]$$

The negative sign in equation 188 indicates that, if the positive direction of flow of charge around the loop is related to the direction of positive flux by the "right-hand rule," a positive increase of flux produces a negative current.

From equation 188, Faraday's law of induction is deduced. The equation can be written

$$RQ = -\Phi \qquad [189]$$

and by differentiation

$$R\frac{\partial Q}{\partial t} = RI = -\frac{\partial \Phi}{\partial t} \qquad [190]$$

It was shown in Chapter V that in a closed circuit the product RI is equal to electromotive force around the circuit, so, from equations 185 and 190,

$$\text{Electromotive force} = \oint \mathbf{E} \cdot d\mathbf{s} = -\frac{\partial \Phi}{\partial t} \qquad [191]$$

This is Faraday's famous law. The flux Φ is that which passes through a surface bounded by the conductor and corresponds to the usual concept of flux linkages (see page 91). Flux is measured in webers and electromotive force in volts.

Magnetically induced electromotive force around a loop of wire is given by equation 191, and it is expressed as the line integral of the electric field along the conducting loop. A generalization of this experimental relation will now be made: It will be assumed that a changing magnetic field induces an electric field according to equation 191, not only in conducting material, but also in non-conducting material and even in empty space. This is reasonable, for, if the loop of copper wire of Experiment VII is replaced by a loop of poorly conducting material such as a nickel-chromium alloy, the induced voltage in the loop remains exactly the same. Voltage is induced even in a piece of wood in a changing magnetic field, and, if the current that flows is very small, it is merely because the resistivity of the wood is quite high. This fact can be established experimentally, if suitably delicate instruments are available,

VOLTAGE INDUCED BY MOTION

and it is not a difficult step to assume that, if this is true in all materials, it is true in air or even in empty space. At first sight this would appear to be a mere quibble, for what does it matter whether an electric field is produced if there is no material present for it to act upon? But it will be seen a little later that this concept is in reality of the utmost importance, and that without electric and magnetic fields producing each other in free space there could be no transmission of radio, light, or other electromagnetic waves through vacuum.

With this assumption (which establishes that there is an induced electric field that is continuous in space), Stokes' theorem can be applied to the integral of equation 191. At the same time expression 187 may be substituted for flux, giving

$$\int (\nabla \times \mathbf{E}) \cdot d\mathbf{a} = -\int \frac{\partial \mathbf{B}}{\partial t} \cdot d\mathbf{a} \qquad [192]$$

These two integrations are both over a surface bounded by the conductor of the experimental apparatus, which may be of any size, shape, or orientation, and the two sides of the equation can be equal under all circumstances only if

$$\nabla \times \mathbf{E} = -\frac{\partial \mathbf{B}}{\partial t} \qquad [193]$$

This equation shows that an electric field will have curl in a region in which the magnetic field is changing with time. The electro*static* field, it will be remembered, in which nothing changes with time, has no curl. The great importance of this equation will appear in Chapter VIII.

Voltage Induced by Motion. In deriving the right-hand member of equation 192 from equation 191, it is assumed that flux through a loop changes only because the magnetic flux density is changing. But it is also possible for the loop to be traveling through space, as, for example, in an electric generator, and its motion through the magnetic field may then alter the amount of flux that passes through the loop even though the strength of the magnetic field at every fixed point of space is constant. In general,[4] if a medium is moving with velocity **v** through

---

[4] See a more advanced treatise on electromagnetic theory, such as *Classical Electricity and Magnetism*, by Abraham and Becker, G. E. Stechert & Co., New York, or *Principles of Electricity and Electromagnetism*, by G. P. Harnwell, McGraw-Hill Book Co., New York, 1938.

space in which the magnetic field is **B**, there will be an electric field induced in the medium, as a result of its motion, equal to $\mathbf{v} \times \mathbf{B}$. This is in addition to any electric field that results from a changing magnetic field strength, and, when it is considered, equation 193 (which is valid for media at rest) becomes (for moving media)

$$\nabla \times \mathbf{E} = -\frac{\partial \mathbf{B}}{\partial t} + \nabla \times (\mathbf{v} \times \mathbf{B}) \qquad [194]$$

EXPERIMENT VIII. A magnetic field has been defined in terms of the force exerted on a wire carrying current. It has also been seen that a changing magnetic field induces an electric field. But neither the source of the magnetic field nor its configuration in space has yet been considered. More experimental information is required for this purpose, and two more experiments will now be described. These are very closely analogous to Experiments II and III that were performed in studying electric fields. It will be seen that both of these experiments in the magnetic field, Experiments VIII and IX, can be done only in a magnetic field that is not changing with time; such a field, by analogy to the electrostatic field, is called *magnetostatic*.

Experiment VIII is performed with an instrument for measuring magnetic flux density. It may be either the current-carrying wire that was used in Experiment VI or the loop and ballistic galvanometer of Experiment VII. The latter would be the more practical. Indeed an exploring loop, usually of many turns wound tightly together, connected by flexible leads to a properly calibrated ballistic galvanometer of extremely long natural period, is a common laboratory instrument known as a "fluxmeter."

The instrument for measuring magnetic flux density is used to determine the normal component of the quantity **B** at all points on a closed surface in a magnetic field. The closed surface is merely an imaginary one, and it may have any shape or size. This experiment must be repeated for very many such surfaces, and the conclusion from the experimental data is that in all cases the summation of the magnetic field over every closed surface is zero. That is,

$$\oint \mathbf{B} \cdot d\mathbf{a} = 0 = \int \nabla \cdot \mathbf{B} \, dv \qquad [195]$$

Applying Gauss's theorem to this experimental result, it appears that

Experiment II $\oint E \cdot ds = 0$ (electrostatic)
III $\epsilon_0 \oint E \cdot da = Q$

THE MAGNETIC FIELD

the magnetic field has no divergence under any circumstances:

$$\nabla \cdot \mathbf{B} = 0 \qquad [196]$$

Lines of magnetic flux are therefore continuous, for it follows from equation 196 that no magnetic flux line has a beginning or end. Every one is a closed loop.

Experiment VIII is performed in material substance of all kinds, as well as in free space, and the result is the same: divergence of the magnetic field is always zero. It is obviously impossible to explore the magnetic field within a solid substance such as brass or iron by measuring force on a conductor, but the fluxmeter loop can be used to determine the total magnetic flux within any piece of solid material, and the result is found to be always consistent with equation 196.

EXPERIMENT IX. This, the last of our experiments, may be done with either of the instruments suggested for Experiment VIII. As before, however, the exploring loop of a fluxmeter is the only way to determine the amount of magnetic flux within solid material. The measuring instrument is used in Experiment IX to determine the magnetic flux density at every point of a closed path, and, by summing the tangential component of the magnetic flux density along the chosen path, the integral $\oint \mathbf{B} \cdot d\mathbf{s}$ is evaluated.

It is first discovered that, if the path of integration lies in homogeneous material, the value of the integral is proportional to the amount of electric current surrounded by the path of integration. If no electric current flows through a surface bounded by the path of integration, the value of the integral (in homogeneous material) is zero. If current flows in such a way as to link the path of integration, however, the value of the integral is given by

$$\frac{1}{\mu}\oint \mathbf{B} \cdot d\mathbf{s} = I \qquad [197]$$

The current is I, and μ is a value that is characteristic of the material. The coefficient μ is called **permeability**. The permeability of empty space [5] is designated by μ_0, and, when I is in amperes, $\mathbf{B}$ in webers per square meter, and $\mathbf{s}$ in square meters, $\mu_0 = 4\pi \times 10^{-7}$ or nearly 1.257×10^{-6}.

[5] The permeability of free space is arbitrarily made to be unity in the unrationalized system of electromagnetic units; this requires that current be measured in abamperes, $\mathbf{B}$ in gausses, $\mathbf{s}$ in centimeters, and that the right-hand member of equation 197 include a factor of 4π. In Gaussian units, the factor in equation 197 is $4\pi/c$, as current is in statamperes.

This experiment may be performed on a ring of iron, as in Fig. 36. The exploring loop of the fluxmeter is wrapped around the iron to measure the flux that exists within the metal. A wire is threaded through the ring and is adjusted to pass normally through its center. A known value of current is then allowed to flow in the wire, and the resulting deflection of the ballistic galvanometer is observed. This is repeated with the exploring loop at various places on the ring (although it is found that the galvanometer reading is the same at all locations). The iron ring is then removed, and the experiment is repeated with the exploring loop of the fluxmeter enclosing merely empty space. Since the readings of the fluxmeter when used on the iron ring are some hundreds

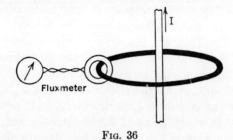

Fig. 36

of times greater than when iron is absent, it follows that the permeability of the iron (under the conditions of the experiment) is several hundred times the permeability of free space.

The permeability of a material, such as iron, is represented by μ. The permeability of free space is μ_0, with the numerical value given above. The *relative* permeability of any material (relative, that is, to free space) is μ/μ_0, and it is this value of relative permeability that is usually found in tables and charts showing the magnetic properties of materials.

Experiment shows that the permeability of most materials is practically that of free space. Only iron, cobalt, nickel, and certain alloys have magnetic permeabilities differing from μ_0 by as much as a few parts in a million. These have high relative permeabilities, ranging up to many thousand, and, because iron is typical of the group, they are known as **ferromagnetic** materials. They are extremely unsatisfactory for analytical study because their permeability is not constant but depends upon the magnetic flux density (as exemplified in the extreme case by magnetic saturation) and, what is worse, the permeability is affected by the previous magnetic history of the material (as seen in the phenomena of hysteresis and permanent magnetism). Fortunately it is not often necessary to consider ferromagnetic materials in connection with electric waves. In this chapter it will be assumed that permeability is constant even in those ferromagnetic materials in which it differs significantly from μ_0.

Finally, when the investigation of Experiment IX is extended to the measurement of magnetic flux density along paths that are partly in one material and partly in another, it is necessary to associate the proper

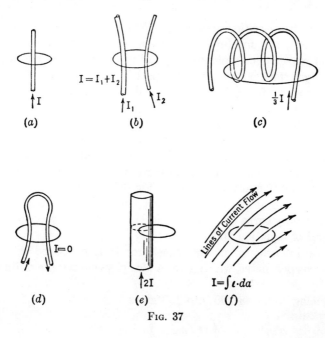

Fig. 37

value of permeability with each part of the path. For a path of integration in non-homogeneous material, equation 197 accordingly becomes

$$\oint \frac{\mathbf{B} \cdot d\mathbf{s}}{\mu} = I \quad [198]$$

By introducing a new symbol **H** representing the vector field of **magnetic intensity**, defined in accordance with the relation $\mathbf{B} = \mu \mathbf{H}$, it is possible to write equation 198 as:

$$\oint \mathbf{H} \cdot d\mathbf{s} = I \quad [199]$$

This equation is found to be true for all possible closed paths of integration, and it sums up the results obtained by performing Experiment IX.

Equation 199 is an equation of **magnetomotive force**. The current I may be the current in a single conductor, as in Fig. 37a, or in several conductors, as in Fig. 37b, c, or d, or in part of a conductor, as in e.

The current may be merely a diffuse flow of charge throughout the entire region, as in f. In any case it may be defined as the integral of the current density over a surface bounded by the path of integration of equation 199:

$$I = \int \iota \cdot da \qquad [200]$$

The surface of integration of equation 200 may be *any* surface bounded by the closed path, for (under the circumstances of the experiment), if a line of current flow links the path of integration, the line will pass through *any* surface bounded by that path.

Magnetomotive force, the left-hand side of equation 199, is directly proportional to current. In practical units, as in equation 199, it is numerically equal to current, and the unit of magnetomotive force is the *ampere* or *ampere-turn*. Dimensionally it is enough to say that magnetomotive force is measured in amperes, but in many practical cases the magnetic effect of a small current is multiplied by winding a wire into a coil of many turns, as in Fig. 37c, and it is customary to speak of magnetomotive force in ampere-turns. But multiplying amperes in a wire by turns in a coil of that wire is merely a way to evaluate the integral of equation 200.

Magnetic intensity or magnetizing force **H** is measured in units of ampere-turns per meter, which gives a nice physical picture of its meaning.

Introducing equation 200 into 199 gives

$$\oint \mathbf{H} \cdot d\mathbf{s} = \int \iota \cdot d\mathbf{a} \qquad [201]$$

By means of Stokes' theorem, the left-hand member of equation 201 can be changed from a line integral around a closed path to a surface integral over a surface bounded by that path:

$$\oint \mathbf{H} \cdot d\mathbf{s} = \int (\nabla \times \mathbf{H}) \cdot d\mathbf{a} = \int \iota \cdot d\mathbf{a} \qquad [202]$$

The second and third members of equation 202 can therefore be integrations over the same surface, which may be any surface, and it follows that at each point

$$\nabla \times \mathbf{H} = \iota \qquad [203]$$

This equation says that the curl of the magnetic field at any point is proportional to the current density at that point. Where there is no current, the field has no curl and is therefore lamellar, but, in space through which current is flowing, the magnetic field has more or less curl. There is curl, for instance, in a magnetic field *within* a wire that is carrying current.

Convention Regarding Sign. In equations 197, 198, and 199, the meaning of the algebraic sign is as yet undefined. Equation 199, for example, states that current flowing in a wire produces magnetomotive force along a path encircling that wire. The question must arise: In which way is the magnetomotive force directed? What is meant by a positive current or a positive magnetomotive force?

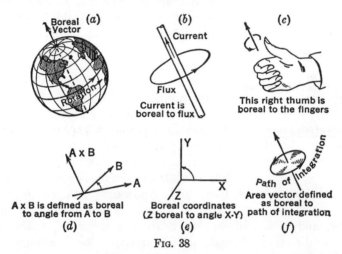

Fig. 38

To clarify this situation a new definition is needed. When direction around a circle is to be related to direction normal to the plane of the circle, one or the other of the possible relations must arbitrarily be accepted as positive. It is customary to assume that, if a circle were drawn in the plane of this page, and if the positive direction around the circle were taken to be counterclockwise, the positive normal direction would be out of the page. If, on the other hand, the positive direction around the circle were taken to be clockwise, the positive normal direction would be in. To express this relation in a single word, let us use the term *boreal*.[6] This makes it possible to state quite simply the accepted convention relating circuital and axial directions: the boreal direction is positive.

---

[6] *Boreal* is derived from the rotation of the earth and signifies a northerly direction compared to the rotation of the earth, or any similar relation between axial direction and rotation. It is from the same root (*boreas*, the north wind) as "aurora borealis." This term is based upon the rotation of the earth as a defining standard, as are Faraday's terms anode and cathode, from west and east. The opposite of *boreal* is *austral*, as in "Australia."

THE MAGNETIC FIELD

When a so-called right-hand screw is turned, it advances in a direction boreal to its rotation. The thumb of the right hand is boreal to the fingers when held as in Fig. 38c. This establishes the "right-hand rule" which is, indeed, the most convenient way to find the boreal direction in individual cases. As in equation 199, current in a wire is boreal to the magnetomotive force it produces; hence, if the right thumb is pointed along the wire in the direction of current flow, the fingers indicate the positive magnetomotive force (see Fig. 38b).

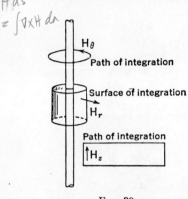

Fig. 39

Equations 201 and 202 relate a surface integral to a line integral, the line integral being taken about the boundary of the surface. Here, again, a convention regarding sign is required. If area is to be considered a vector quantity, represented by a vector normal to the surface, it must be known which sense the vector is to have. In previous chapters it has been enough to say that the sense of the vector was outward from a closed surface; but in equations 201 and 202 the surface is not a closed surface, and "outward" has no meaning.

When a surface is not closed, it has a boundary, and it is sufficient to relate the direction of the area vector to direction around the boundary; the accepted convention is that the boreal direction is positive. Thus in equation 201 the line integral of the left-hand member may be taken in either direction, selected arbitrarily, and this choice of direction around the boundary defines also, by the above convention, the positive sense of area in the right-hand member of the equation. With this convention all ambiguity is removed.

Example. As an example of a solution for a magnetic field, let us compute the field in air about a long straight wire of circular cross section, as in Fig. 37a or 39. The wire is carrying current I, and the three components of the magnetic field are to be found.

Equation 199 gives the integral of the magnetic field along any closed path surrounding the wire. First, to apply this equation, let us select a circular path of integration that is concentric with the wire as in Fig. 39. Along this path, **H** is constant; this must be true because of symmetry, for one point on this circle about a circular conductor cannot be distinguished from another.

Call the component of **H** that is tangent to this circular path of integration H_θ. Only this component will contribute to the scalar product of equation 199, so

$$\oint H_\theta \, ds = I \qquad [204]$$

But since, at any constant radius r, H_θ is constant, this becomes

$$H_\theta \oint ds = I \qquad [205]$$

The integral of ds is merely the length of the circular path of integration, and at radius r

$$H_\theta 2\pi r = I \qquad [206]$$

from which

$$H_\theta = \frac{I}{2\pi r} \qquad [207]$$

Call the component of **H** that is parallel to the axis of the wire H_z. If such a component exists, it must, because of symmetry, be equal at all points that are equal radial distances from the wire.

Then call the component of **H** that is in a radial direction H_r. Consider a surface around the conductor and coaxial with it, the general shape of a tin can or a round pillbox. That is, the surface is a closed surface, composed of a cylindrical section and two circular plane sections as in Fig. 39. It is desired to obtain $\int \mathbf{B} \cdot d\mathbf{a}$ over this surface for use in equation 195. Since H_θ does not intersect this surface it contributes nothing to $\int \mathbf{B} \cdot d\mathbf{a}$. The axial component H_z contributes nothing to $\int \mathbf{B} \cdot d\mathbf{a}$, for, being equal at equal radial distances from the conductor, if it adds to the integral over one plane surface, it subtracts an equal amount over the other, and it does not intersect the cylindrical surface. If the radial component H_r existed, however, it would give $\int \mathbf{B} \cdot d\mathbf{a}$ over this tin-can-like surface some value different from zero. Since equation 195 says $\int \mathbf{B} \cdot d\mathbf{a}$ is always zero, it follows, when all possible tin-can-like surfaces are considered, that H_r must be zero everywhere.

It remains to evaluate H_z. Consider a line of integration as indicated in Fig. 39. It is rectangular in shape. One of the sides is parallel to the axis of the wire and is fairly close to the wire; the other parallel side is unlimitedly far away. When integrating **H** · $d\mathbf{s}$ around this rectangle.

nothing is contributed to the integral by H_θ, which is everywhere normal to the path. It has been shown that $H_r = 0$. There remains only H_z, which will (if it exists) contribute to the integral along the short sides of the path.

The total integral around the path is zero, for the path links no current (equation 199). The contribution of H_z to the integral must therefore be zero, and this is possible only if H_z has the same value near the conductor that it has at an unlimited distance, or if $H_z = 0$. Since it is impossible that a conductor carrying finite current should produce a uniform magnetic field through infinite space (for this would require infinite energy), it must be concluded that H_z, as well as H_r, is zero.

Finally, therefore, we determine that only H_θ exists, and its value is given by equation 207, which is known as the *Biot-Savart* law. The solution of this simple problem has been carried out in great detail because it illustrates the use of special paths of integration for reaching conclusions regarding magnetic and electric fields.

Force Between Currents. In Experiment VI it was found that there is a mechanical force on a conductor that carries current in the neighborhood of another conductor also carrying current. The effect must be mutual, and each current exerts a force on the other. The amount of this force can now be determined.

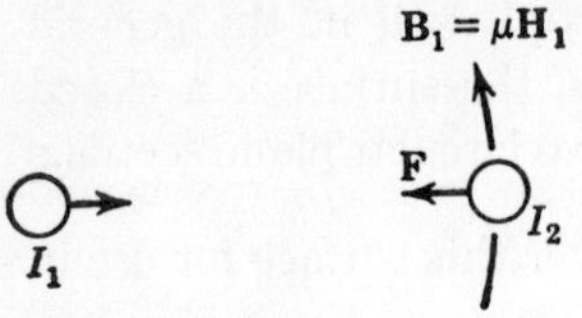

The determination of force between parallel wires is simplest and will be illustrated here. The same method can be applied, if desired, to the general case of any conductors.

Fig. 40

Figure 40 shows a cross section of two conductors. Currents I_1 and I_2, measured in *amperes*, are flowing in the conductors, both being directed out of the page. If the distance in *meters* between the two conductors is d, the magnetic field produced by I_1 at the distance of conductor 2 is (by equation 207)

$$B_1 = \mu H_1 = \frac{\mu I_1}{2\pi d} \qquad [208]$$

The direction of this field is normal to a line connecting the two conductors, as shown in the figure. There is a mechanical force on conductor 2 as given by equation 186, and, when equation 208 is substituted into equation 186, it is seen that the magnitude of the force is

$$F = I_2 L_2 \frac{\mu I_1}{2\pi d} = \frac{\mu}{2\pi} \frac{I_1 I_2}{d} L_2 \qquad [209]$$

or the force in newtons per meter length of conductor 2 is

$$\frac{F}{L_2} = \frac{\mu}{2\pi} \frac{I_1 I_2}{d} \qquad [210]$$

This is a scalar equation. It gives magnitude of force only. The vector product **L** × **B** in equation 186 is here equal in magnitude to the product of the scalar magnitudes L and B, because **L** and **B** are normal to each other; the direction of force is given by the direction of this vector product, and, since, in Fig. 40, **L** is out of the page (corresponding to the direction of flow of current), and **B** is up, **F**, being boreal to the angle from **L** to **B**, is directed toward conductor 1.

It is evident that equation 210 gives also the magnitude of the force exerted on conductor 1 by the current in conductor 2. The direction of such a force is toward conductor 2. Therefore we have determined that two conductors carrying current in the same direction attract each other, the amount of force being given by equation 210. This is a simple form of **Ampère's law.** As given here it assumes that the spacing between conductors is large compared to the diameter of either conductor and that the conductors are straight and parallel for an unlimited distance.

If either current were reversed in direction, changing the sign of I in equation 186, the direction of the force between conductors would be reversed and would become repulsive. But, if both currents were reversed, there would again be attraction. From this has arisen the easily remembered but somewhat loose statement that "like currents attract, unlike currents repel."

Magnetic Flux Linkages. Current flowing in a coil of wire, as in Fig. 37c, produces a magnetic field. The configuration of the field is such that flux lines extend axially through the coil and return, rather widely dispersed, in outer space. Each flux line thus passes at least once, and possibly several times, through a surface bounded by the conductor. Such a surface in Fig. 37c can be visualized as a sheet of rubber with its edge attached to the conductor. When the wire is bent into a helix, the rubber sheet is stretched into a complicated shape that can be imagined more readily than it can be drawn or described. The closed line shown in Fig. 37c might represent a flux line; such a line passes three times through the surface bounded by the conductor. It will be recognized that each penetration of this surface by the flux line is equivalent to the ordinary concept of a "flux linkage."

Magnetic Potential. In the discussion of electrostatic fields, a good deal of attention was given to the potential field V. This is a scalar field. It is known to exist because $\nabla \times \mathbf{E} = 0$, and its gradient (with negative sign) is $\mathbf{E}$.

Now one naturally inquires whether there is a similar magnetic scalar potential field. Equation 203 is enough to give warning that such a field may not exist, for $\nabla \times \mathbf{H}$ is not always zero. This much can be said: in any region where there is no current flowing, the (static) magnetic field has no curl, and a scalar magnetic potential exists. It is sometimes useful, but not in connection with waves, and so it will not be discussed here.

Vector potential was mentioned in Chapter III. No effort was made to find an electrostatic vector potential, for the divergence of the electrostatic field is not always zero, and where it is not the vector potential field will not exist. To be sure, an electrostatic vector potential might be found in space that contains no charge, for there divergence is zero. The magnetic field, on the other hand, has zero divergence everywhere (equation 196), and it is possible to find a magnetic vector potential that turns out to be of real value.

Let us call the magnetic vector potential $\mathbf{A}$. By definition,[7] then, $\mathbf{A}$ is such a field that

$$\mathbf{H} = \nabla \times \mathbf{A} \quad [211]$$

Since $\iota = \nabla \times \mathbf{H}$, from equation 203, it follows that

$$\iota = \nabla \times \nabla \times \mathbf{A} \quad [212]$$

By using the identity given in footnote 5, page 32, this becomes

$$\iota = \nabla(\nabla \cdot \mathbf{A}) - \nabla^2 \mathbf{A} \quad [213]$$

Equation 213 is somewhat similar to Poisson's equation for electrostatics and gives a means of finding current distribution if the magnetic vector potential field $\mathbf{A}$ is known. However, $\mathbf{A}$ is not known; it is not yet even fully defined. All that is known is the curl of $\mathbf{A}$. Any number of fields may have the same curl, but only one field can have a given curl and also a given divergence.[8] Let us therefore specify the divergence in some convenient manner, and $\mathbf{A}$ will be fully defined. For

[7] Many authors define $\mathbf{B} = \nabla \times \mathbf{A}$. These definitions of A differ by a factor μ, and either is satisfactory if used consistently.

[8] Boundary conditions as well as divergence and curl must be known for the field to be completely defined. It is here sufficient to know that the vector-potential field vanishes at infinity. There is a theorem, known as the theorem of uniqueness, that says, "A vector field is uniquely determined if the divergence and curl are specified, and if the normal component of the vector is known over a closed surface, or if the field vanishes (at least as rapidly as $1/r^2$) at infinity."

the magnetostatic field the most convenient choice is the obvious one. let us say that

$$\nabla \cdot \mathbf{A} = 0 \qquad [214]$$

Then equation 213 is simplified to

$$\nabla^2 \mathbf{A} = -\iota \qquad [215]$$

and this is quite analogous to Poisson's equation

$$\nabla^2 V = -\frac{\rho}{\epsilon} \qquad [118]$$

If current is known and the magnetic vector potential is to be found, equation 215 is a differential equation for which a solution must be sought. It was shown in Chapter IV that Poisson's equation has an integral solution:

$$V = \frac{1}{4\pi\epsilon} \int \frac{\rho\, dv}{r} \qquad [146]$$

and this helps to find the solution for the magnetic vector potential.

Equation 215 is a vector equation; when it is expanded into three scalar equations, each one is seen to be identical in form to equation 118. The mathematical solution of each must therefore be formally identical with 146, with only a change of symbols. These component equations and their solutions are as follows:

$$\nabla^2 A_x = -\iota_x \qquad A_x = \frac{1}{4\pi} \int \frac{\iota_x\, dv}{r}$$

$$\nabla^2 A_y = -\iota_y \qquad A_y = \frac{1}{4\pi} \int \frac{\iota_y\, dv}{r} \qquad [216]$$

$$\nabla^2 A_z = -\iota_z \qquad A_z = \frac{1}{4\pi} \int \frac{\iota_z\, dv}{r}$$

A_x, A_y, and A_z then combine to give

$$\mathbf{A} = \frac{1}{4\pi} \int \frac{\iota}{r}\, dv \qquad [217]$$

The interpretation of equation 217 is similar to that of equation 146. Vector potential is found by integrating over as much of space as may be carrying current, r being the distance from each elementary unit of current to the point at which vector potential is being determined. If current is flowing in a circuit, the integration need only be performed

94   THE MAGNETIC FIELD

about the circuit; elsewhere, where there is no current density, the contribution to the integral is zero. The result of the integration is the vector-potential field.

It will be seen that current flowing in the x direction produces only an x component of vector potential. In general, the direction of the vector potential is the same as that of the element of current by which it is produced. Also, the vector-potential field is strongest near the current that produces it and fades away gradually at greater distances. The vector-potential field is sometimes described as "like the current distribution but fuzzy around the edges," or "like a picture of the current out of focus." These inelegant ideas are distinctly helpful. It is interesting to consider how they apply to the example at the end of Chapter III (page 47).

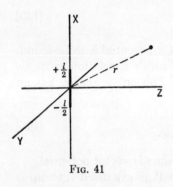

Fig. 41

Computing a magnetic vector potential from equation 217 is as difficult as computing electrostatic scalar potential from equation 146. It has such important applications, however, that an example will be given for which an approximate solution is found.

Example. Let us consider a short length of wire carrying alternating current

$$i = I \sin \omega t \qquad [218]$$

The wire is isolated in space. How current is supplied to it is a problem that need not concern us at the moment (but it can be visualized as part of a circuit). The frequency of alternation of the current is so low that the equations of magnetostatics apply, and we are thus considering what is known as the quasi-stationary state. In other words, our little antenna does not radiate.

The length of the wire is l, and we locate coordinates in such a way that the conductor extends from $-l/2$ to $+l/2$, as in Fig. 41.

We want to use equation 217 to find the magnetic vector potential field about the wire. The only current in which we are interested is that flowing in the wire, so the integration is performed along the wire only. Since current flows only in the x direction, there is only an x component of vector potential. If the wire has small cross section da, the element of current is merely $i\,dx$, instead of $\iota_x\,dv$, for $\iota_x\,dv = \iota_x\,da\,dx = i\,dx$. Then

$$A_x = \frac{1}{4\pi} \int_{-l/2}^{l/2} \frac{I \sin \omega t}{r} dx \qquad [219]$$

The exact solution of this would be difficult, for strictly speaking the quantity r is the distance, to a point in space, from a point on the wire that moves as the integration proceeds, so that r is a function of x. But, if potential is being determined at a distance from the wire that is many times greater than the length of the wire, making $r \gg l$, then r may be measured from the middle of the wire as in Fig. 41 with very little error. This approximation permits integrating with r constant, and the vector potential is simply

$$A_x = \frac{I \sin \omega t}{4\pi r} \int_{-l/2}^{l/2} dx = \frac{Il}{4\pi r} \sin \omega t = \frac{il}{4\pi r} \qquad [220]$$

This is the magnetic vector potential (the other two components of **A** being zero). **H** can be found by computing the curl, by equation 211. It must be remembered that this solution is not accurate quite close to the wire that carries current. (It is instructive to note that **A** of equation 220 has divergence, contrary to equation 214. This is because the wire of Fig. 41 is only part of a circuit and equation 175 is not satisfied. A solution that satisfies equation 175 is on page 165.)

Magnetic Energy. Energy is required when the magnetic field is produced. The energy comes from the electric circuit, as follows. When current starts to flow in a circuit, it produces a magnetic field (equation 198). As the magnetic field grows it induces an electric field in and near the region of the magnetic field; the integral of this electric field along the circuit is electromotive force (equation 191). An increasing magnetic field induces electromotive force in the circuit in such a direction that it opposes the increase of current.[9] This induced electromotive force must be overcome by the applied voltage. It is readily shown that the product of current, voltage, and time is energy. Thus energy is taken from the circuit as the magnetic field is produced. An equal amount of energy is returned to the circuit when current ceases to flow, as the magnetic field dies away.

It is assumed that energy taken from the electric circuit during the formation of a magnetic field is stored throughout that field. If magnetic energy density is taken to be $\frac{1}{2}\mathbf{B} \cdot \mathbf{H}$ or $\frac{1}{2}\mu H^2$, total energy existing in a magnetic field is

$$\text{Magnetic energy} = \frac{1}{2} \int \mathbf{B} \cdot \mathbf{H}\, dv \qquad [221]$$

Since it can be shown that this is equal to the total energy required to establish a magnetic field, the assumed value of energy density is justified.

[9] Thus Lenz's law is deduced from equations 191 and 198.

The problem is analogous to that of electrostatic energy in Chapter IV, and it will be seen by comparison with equation 157 that the expression for magnetic energy is quite analogous to that for electrostatic energy.

Theories. There are two usual points of view for considering the magnetic behavior of materials. One considers the magnetic intensity **H**, produced by electric current or by a permanent magnet, to be a kind of magnetic driving force, and **B** is considered the resulting magnetic flux density. Permeability, then, is a measure of the ease with which flux can be produced in a given material. Consistent with this concept, the line integral of **H** is called magnetomotive force. Magnetomotive force is considered to be analogous to electromotive force in an electric circuit, in which case **B** is analogous to current density and flux to current.

Because of analogies to familiar concepts, this view of magnetic behavior is very convenient for visualizing magnetic fields and quite useful in computation. But, for theoretical purposes and for gaining an understanding of the physical processes underlying magnetic behavior, an entirely different concept is generally accepted.

It is believed as a physical theory that the field **B** is not dependent on the nature of material but is determined by the flow of electric current, and nothing else. The introduction of an unexplained factor called "relative permeability" is thus avoided. But this concept is valid only if *all currents* are taken into account. This includes currents *within* the atomic structure of matter, as well as ordinary currents which are carried by electrons between the atoms of conducting material. Let us consider iron first, because it is an extreme case.

It is believed that material is made of atoms, and that each atom consists of a nucleus with electrons about it. Because of electronic rotations and spins, which constitute circulating currents within the atom, some atoms are equivalent to small loops carrying current and will produce magnetic fields. This is true in paramagnetic materials with ferromagnetic substances as extreme examples.

In ordinary iron the many atoms are oriented at random, and, although each atom is a small circuit forever carrying current, a piece of iron containing a great number of atoms is on the whole not a magnet. However, if it is placed in an external magnetic field, there is a force on each atom that tends to orient all atoms the same way. Atoms of iron are able to respond to this force; in a magnetic field they reorient themselves, apparently in small groups, and become aligned with the external field. Then, in an extreme case of polarization, all the many subatomic electric currents will cooperate in strengthening the electric field that caused their orientation. This results in a total magnetic field that is tremendously stronger than can be accounted for by electric current in the external circuit alone, which serves mainly to orient the iron atoms.

Diamagnetic Materials. A word must be said about diamagnetic materials, although diamagnetism is an extremely slight effect. It is supposed to result from what may be considered induced currents within the atoms of material.

Figure 42 shows a coil of wire to which a battery may be connected, and a ring of highly conductive material. When the battery is con-

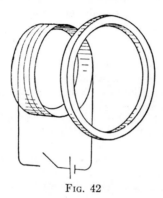

Fig. 42

nected to the coil, a magnetic field is produced, part of which links with the ring. Formation of such a field through the ring induces electromotive force in the ring, and current flows; the result of this induced current in the ring is to weaken the magnetic field produced by the coil. If the material of the ring had perfect conductivity, current induced in the ring would flow as long as there was current in the coil, and the total magnetic field would be always less intense than it would have been in the absence of the ring.

It is believed that electrons within atoms of material substance act like the ring of Fig. 42. When a magnetic field is produced, their motion will be altered, producing the equivalent of a demagnetizing current. Hence the magnetic field in the material is weaker than it would have been in free space, and the material is diamagnetic.

This effect is quite independent of the orientation of polar atoms which constitutes paramagnetism. In fact, it is supposed that all materials are diamagnetic but that some have also a paramagnetic tendency, and that the latter effect is in many cases more marked than the former, with the result that the material is on the whole paramagnetic or even ferromagnetic.

98 THE MAGNETIC FIELD

The demagnetizing circulating current of diamagnetism is always extremely small, as it affects the magnetic field by only a few parts in a million.

PROBLEMS

1. Would it be possible to *define* magnetic field strength as a vector *parallel* to the magnetic force on the exploring wire of Fig. 34?

2. What is the direction of the magnetic field in Fig. 34 if current is flowing downward in the exploring wire? Check with equation 186.

3. If magnetic flux links a coil of wire of several turns, the voltage induced in the coil by a change of flux is proportional to the number of turns. How is this expressed by equation 191? What is the surface of integration for that equation in such a case?

4. There are ten turns of wire wound in a layer on a wooden spool, and a current of 5 amperes flows in the wire. Ten more turns are then wound close upon the first layer, and current is passed through all turns in the same direction around the spool. With the added turns, current is reduced to 4 amperes. Using equation 201 with a path of integration passing axially through the spool, find how much the magnetic field near the center of the spool is changed. Draw a sketch showing the direction of current, the direction of flux, and the path of integration of equation 201.

5. Show that the magnetic field at a radius r within a copper conductor carrying current I is $(Ir)/(2\pi r_0^2)$. The radius of the cylindrical conductor is r_0, which is of course greater than r. Current is uniformly distributed across the conductor cross section.

6. Find the curl of the magnetic field of Problem 5. Does it agree with equation 203?

7. How is the magnetizing current of a transformer related to the number of turns of the primary winding, if there is no change of the general dimensions of the transformer or of the applied alternating voltage? From which equations do you reach this result?

8. Find and plot the magnitude of magnetic vector potential $\mathbf{A}$ along a radial line passing through the center of the conductor of Problem 5. Show, in a single curve, the intensity of the vector potential both within and outside the conductor. The following conditions are to be met: $\nabla \times \mathbf{A} = \mathbf{H}$; $\nabla \cdot \mathbf{A} = 0$; when $r = 0$, $\mathbf{A} = 0$; there is no discontinuity in $\mathbf{A}$ at $r = r_0$.

9. Compute (from the vector-potential field of equation 220) the magnetic field about a short wire carrying alternating current. Spherical coordinates are suggested; see Table II and Fig. 58.

10. Draw arrows in the X-Z plane of Fig. 41 to show $\mathbf{A}$; use dots and crosses to show $\mathbf{H}$.

11. Compare the result of Problem 9 with the magnetic field of an oscillating doublet (or dipole) as given in electrical physics books.

CHAPTER VII

Examples and Interpretation

Most of the physical relations that have been discussed in the previous chapters are familiar. They are simple laws of electrostatics, magnetic flux, the steady flow of current, and induced voltage. If they have appeared strange it is because they have been generalized to apply to the broadest possible range of conditions. These generalized relationships have something in common with disembodied spirits and seem unsubstantial to most of us until they are attached to concrete situations. The unaccustomed notation of vector analysis has done nothing to relieve this situation, although it has done a great deal to save us from wandering in a maze of differential equations.

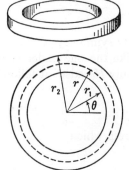

Fig. 43

The purpose of this chapter is to supply a few concrete illustrations. There will be no new experimental evidence.

Example 1. A solenoidal coil of many turns of fine wire is wound on a wooden core, the shape of which is shown in Fig. 43. The core is a ring of rectangular cross section. There are N turns of wire wound upon it, each carrying current I. It is desired to find the magnetic field produced by the current, and the inductance of the coil.

Cylindrical coordinates may best be used for reference, with the Z axis coinciding with the axis of the ring. The radius of the inner surface of the ring is r_1 and of the outer surface r_2.

From the symmetrical arrangement of the current and from the known nature of the magnetic field, it is apparent that the magnetic field in this case will be circular. That is, H_θ will exist, but H_r and H_z will be zero.[1] This could be proved, of course, but time will be saved if it is accepted without proof.

---

[1] If there is a single layer of turns of wire wound on the core, so that the current follows once around the core while spiraling through the winding, there will be a small component of field (H_z) passing vertically through the space within the core. If a double-layer winding is used this may be completely eliminated, and in any coil it may usually be neglected.

To find the strength of the field, consider a circle of radius r (shown by the dash circle in Fig. 43), to be the path of integration for equation 201:

$$\oint \mathbf{H} \cdot d\mathbf{s} = \int \mathbf{\iota} \cdot d\mathbf{a} \qquad [201]$$

With the line integral on the left-hand side of this equation taken around the circle of radius r, the right-hand member gives the total current passing through the space within the circle. If r is less than r_1 and lies within the ring, or if r is greater than r_2, or if the circular path of integration lies above the ring or below it, no current passes within the circle. In such a case $\mathbf{H}$ is zero, and there is no magnetic field in these regions.

But, if the circle lies within the coil as shown in the figure, the current I passes N times through any surface bounded by the circle, and

$$\oint \mathbf{H} \cdot d\mathbf{s} = NI \qquad [222]$$

By symmetry, the value of $\mathbf{H}$ is constant along a circular path that is, like the one under consideration, concentric with the ring, and, since only H_θ exists, it follows that

$$H_\theta \oint ds = 2\pi r H_\theta = NI \qquad [223]$$

from which

$$H_\theta = \frac{NI}{2\pi r} \qquad [224]$$

According to this equation the magnetic field is not uniform within the coil, but is stronger nearer the inner surface, the field strength being inversely proportional to the radius.

It may be recognized that $\int \mathbf{H} \cdot d\mathbf{s}$ is **magnetic potential difference** (magnetic scalar potential), analogous to electric potential difference or voltage. The magnetic potential difference around a closed path, as in equation 222, is commonly given the name **magnetomotive force** (see page 85).

Let us compute the amount of flux within the solenoidal coil. To find the total flux we integrate $\mathbf{B}$ over the cross section of the core. Since $\mathbf{B} = \mu \mathbf{H}$, the desired value is

$$\Phi = \int \mathbf{B} \cdot d\mathbf{a} = \int \mu H_\theta \, da \qquad [225]$$

If the thickness of the core parallel to the Z axis is z_1,

$$\Phi = \int_{r_1}^{r_2} \frac{\mu NI}{2\pi r} z_1 \, dr = \frac{\mu N I z_1}{2\pi} \ln \frac{r_2}{r_1} \qquad [226]$$

Now the inductance of the coil can be determined. The **inductance** is, by definition,

$$L = \frac{N\Phi}{I} \qquad [227]$$

Φ is the flux that passes through a single turn of the conductor and is

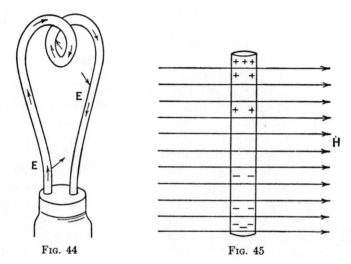

Fig. 44   Fig. 45

therefore the flux in the core. Substituting equation 226 into 227, the inductance of the toroidal coil is

$$L = \frac{N\Phi}{I} = \frac{\mu N^2 z_1}{2\pi} \ln \frac{r_2}{r_1} \qquad [228]$$

This inductance is in henrys, and dimensions are in meters; since the core is wood, $\mu = 1.26 \times 10^{-6}$. The symbol ln indicates the natural logarithm.

Example 2. Figure 44 shows an old-fashioned carbon filament for an electric light. The thickness of the filament is exaggerated so that arrows may be drawn to show the electric field within the filament. A steady current is flowing through the filament, and it is desired to find the electric field that exists.

Within the filament there is uniform current density, flowing everywhere parallel to the surface of the conducting filament, and, since electric field strength is proportional to current density (equation 165), the electric field is also directed along the filament and is uniform in magnitude. If the filament is 10 inches long and the applied voltage is 110 volts, the electric field strength within the filament is everywhere 11 volts per inch. This field is in no way affected by turns and twists of the filament, and the same situation exists in any wire of uniform cross section that is carrying steady current. In a conductor that is not carrying current, there is no electric field.

This last statement might seem contradictory to the fact that an electric field may be induced in a conducting rod that is not part of a circuit and hence cannot carry a steady flow of current. But let us suppose a copper rod, as shown in Fig. 45, is in a changing magnetic field that induces an electromotive force upward in the rod. At the instant the magnetic field begins to induce electromotive force, there will be an electric field in the copper, and for an instant current will flow. Almost at once, however, enough positive charge will accumulate at the top of the rod, and enough negative charge at the bottom, to produce a field equal and opposite to the induced field, and the charge will distribute itself in exactly the right manner to give zero resultant electric field within the copper and there will then be zero current. Although there is no field within the rod in such a case, it does not follow that there is zero electric field elsewhere. There will, indeed, be electric flux emanating from the top of the rod and returning at the bottom, and a voltmeter (if its leads were not in the changing magnetic field) would indicate the value of the induced electromotive force. See page 73 for a discussion of electromotive force, of which this rod is an example.

Returning to consideration of the electric-light filament, there will be electric field also in the space about the filament. There is a potential difference of 110 volts between the two ends of the filament, so there must be electric field in the intervening space. This emanates from charge located on the surface of the filament; charge that was driven at the instant voltage was applied to the filament by a component of field within the filament normal to the filament surface. When this charge reached its final position in proper amount, the normal component of field *within* the filament was reduced to zero, and thereafter the surface charge remained constant.

The electric field in space around the filament is quite complicated. There is a component normal to the surface, due to the surface charge. There is a component tangential to the surface, equal to the field within the filament. The geometry of the filament is not simple, and we will not attempt any quantitative solution of this problem, but merely note that the line integral of **E** · d**s** from any point on the filament to another point on the filament, following any path, must be equal to the potential difference between those two points, and the electric field strength will be distributed so as to make this true. The electric field in space near the surface of the filament, particularly at the ends of the filament, may have an intensity of thousands of volts per inch.

Example 3. When current is changed in the toroidal coil of Example 1 (Fig. 43), the magnetic field is changed proportionately, and the changing magnetic field induces an electric field in and around the toroid. Equation 193 tells us that the induced electric field has curl within the core of the toroid, the amount of curl being

$$\nabla \times \mathbf{E} = -\frac{\partial \mathbf{B}}{\partial t} \qquad [229]$$

Integrating each side of this equation over the cross-section area of the coil, and applying Stokes' theorem to the left-hand member, gives

$$\int \nabla \times \mathbf{E} \cdot d\mathbf{a} = \oint \mathbf{E} \cdot d\mathbf{s} = -\frac{\partial}{\partial t} \int \mathbf{B} \cdot d\mathbf{a} \qquad [230]$$

The second member of this equation is a line integral of electric field. For our purposes we select a path of integration that surrounds the core of the solenoid; it may be practically identical with one turn of the winding that is wrapped upon the core. The integral of induced electric field along such a path is the electromotive force induced in one turn of the coil (as in equation 191); and there are N turns. The surface integral in the right-hand member of equation 230 is, from equation 187, the magnetic flux through the core. So the total electromotive force induced in all N turns is

$$\text{Electromotive force} = -N\frac{\partial \Phi}{\partial t} \qquad [231]$$

This is a familiar equation. Another familiar form results when the definition of inductance given in equation 228 is substituted into equation 231, giving

$$\text{Electromotive force} = -L\frac{\partial I}{\partial t} \qquad [232]$$

The negative sign indicates that, if the current is increasing so that the rate of change of current is positive, the induced electromotive force is negative and opposes the flow of current.

Example 4. When a magnetic field is changing, it induces an electric field even in empty space, as discussed in Chapter VI. The electric field so induced is capable of exerting force on any charged particles that exist in the region of changing magnetic field, and this principle is used [2] in the "induction accelerator" to drive electrons for "atom-smashing" purposes.

Consider a narrow evacuated space between the large, round, flat faces of iron magnetic poles. The evacuated space is rather disk-like, thin from top to bottom, but large in diameter, and a strong magnetic field passes through the thickness of it from the magnetic pole face above to the magnetic pole face below. The magnetic field is produced

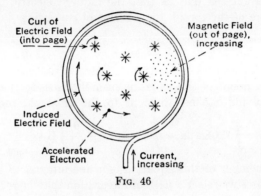

FIG. 46

by current flowing in many turns of wire about the iron field structure (see Fig. 46).

Now consider that current through the winding can be increased from zero to some given value in a very small fraction of a second (1/600 second); the magnetic field in the evacuated space will increase at a correspondingly rapid rate. This will induce an electric field in the space, and electrons or other charged particles released into the evacuated space will have force exerted on them and will be accelerated in a more or less spiral path, gaining velocity.

As a first approximation we may consider that the magnetic field is uniform in the region directly between pole faces, falling suddenly to zero as one passes out from between the iron plates. This is not truly correct, for it is impossible to have an abrupt change from a region of magnetic field to a region of no magnetic field; an abrupt change of magnetic field strength implies infinite curl in the magnetic field (consider the "curl-meter" of Chapter II). In the actual case of magnetic field between pole faces, there will be, as is well known, "fringing" of the lines of force, allowing the field strength to decrease gradually to

[2] "Acceleration of Electrons by Magnetic Induction," D. W. Kerst, *Physical Review*, Volume 60, July 1, 1941, pages 47–53.

zero without curl. (As a matter of fact, the actual accelerator, or "betatron," is designed with a non-uniform field, even between pole faces, to provide stability of electron orbits, but this non-uniformity will not be considered here.)

Where the magnetic field is changing with time, there will be curl of the electric field (equation 193), and the next step of the problem is to find the electric field. When current in the winding is changed, there is a change of magnetic field in the space between pole faces, and, since the field is increasing in intensity without changing in direction, it follows that the rate of change is in the same direction as the field itself. The curl of **E** is, therefore, by equation 193, also uniform and in the opposite direction. The problem is to find an electric field that has such a curl in the circular region between magnets and no curl elsewhere.

Visualize a battery of paddle-wheel "curl-meters" as in Fig. 46, all turning at the same speed and in the same direction. It is apparent that their rotation may be caused by an electric field revolving as a whirlpool with the greatest intensity at the circumference. It will be well to be definite regarding direction.

If an increasing magnetic field is produced by a counterclockwise current in the field winding, as indicated, the magnetic field, being boreal to the current, will be out of the page. Because of the negative sign in equation 193, the curl of the electric field will be into the page, and the circulation of the electric field, being boreal to its curl, will be clockwise. Note that it is this same clockwise field that, induced in the conductor of the field winding, opposes the increase of current according to Lenz's law.

Assuming cylindrical coordinates with the Z axis coinciding with the axis of the magnetic poles, the electric field is in the negative θ direction. Let us assume [3] it to be proportional to r:

$$E_\theta = -Ar \qquad [233]$$

The value of A is to be determined; but it is not a function of r, θ, or z.

From Table II, the curl of the electric field is

$$\nabla \times \mathbf{E} = -\mathbf{k}\left(A + \frac{Ar}{r}\right) = -\mathbf{k}2A \qquad [234]$$

and, from equation 193,

$$\frac{\partial \mathbf{B}}{\partial t} = \mathbf{k}2A \quad \text{or} \quad \frac{\partial B}{\partial t} = 2A \qquad [235]$$

---

[3] See the similar example, more fully worked out, on page 47.

This evaluates A for use in equation 233, and it follows that between the magnetic pole pieces

$$E_\theta = -\frac{r}{2}\frac{\partial B}{\partial t} \qquad [236]$$

Although proportionality between field strength and radius was merely assumed in equation 233, the correctness of the assumption is proved by showing that equation 193 is satisfied by the electric field of equation 236 and the given magnetic field of uniform distribution.

The induced electric field has its maximum value at a radius r_0 equal to the outer radius of the magnetic field. The electric field does not then cease abruptly but dies away at radii greater than r_0 in such a manner that the electric field has no curl. (This would be but slightly modified if fringing of the magnetic field were considered.) To determine the electric field that lies in outer space beyond the magnetic field, it is necessary to find a field that (1) has no curl, (2) is continuous with the field of equation 236 at radius r_0, and (3) vanishes at infinite radius. Conditions (1) and (3) are satisfied by

$$E_\theta = -\frac{A'}{r} \qquad [237]$$

At radius r_0, this must be equal to equation 236, so

$$\frac{A'}{r_0} = \frac{r_0}{2}\frac{\partial B}{\partial t} \qquad [238]$$

From this the value of A' can be determined, and, in the outer region

$$E_\theta = -\frac{r_0^2}{2r}\frac{\partial B}{\partial t} \qquad [239]$$

Equation 239 follows from an assumption of unlimited empty space. In fact, of course, the magnetic field structure and other apparatus must interfere with this ideal condition, and equation 239 is a more or less accurate approximation of the electric field at radii that are not too much greater than r_0.

It is very interesting to compare the electric fields that, in this example, result from a changing magnetic field with the magnetic fields that were shown in Chapter VI to result from current in a long straight wire. The field distributions in and around a long conductor are strictly analogous to those of equations 236 and 239 respectively.[4]

[4] The reason for the analogy is simply that the two field distributions are parallel solutions of the two Maxwellian equations.

Equations 236 and 239 are not the only possible solutions of equation 193 within the regions under consideration. In fact any electric field without curl could be added to these solutions, and the sum would also be a solution of the equation. (A field without curl is, in this case, analogous to a constant of integration.) But there simply *is* no field that is without curl and without divergence and that becomes zero at an infinite distance and is nowhere of infinite strength. Within these obvious physical limitations, our solutions are the only ones that satisfy both the electromagnetic equations and the boundary conditions.

One factor, however, has been neglected. It is safe to do so in this case, for its quantitative importance is insignificant. But in other examples it will be found to be the only important part of the solution. It is **radiation**. When the changing magnetic field produces an electric field about the magnet, a wave of electromagnetic energy travels outward from the apparatus. It carries away from the "induction accelerator" of the present example so small a fraction of the total energy that it is entirely negligible.

But radiation is not always negligible; in radio communication it is the *sine qua non* of practical value. The electromagnetic theory that has been developed in the preceding chapters fails to account for radiation. It deals with electric fields that are **quasi-stationary**—fields, that is, that change so slowly that at any instant they may be regarded as electrostatic.

We have gone farther in magnetics, for we have discovered that the rate of change of a magnetic field is important. A changing magnetic field is capable of producing an electric field. We owe to James Clerk Maxwell the idea that a changing electric field is likewise capable of producing a magnetic field. With the consideration of this additional hypothesis we advance from the quasi-stationary state to the **electrodynamic** state.

PROBLEMS

1. A coil of many turns of fine wire is wound on a long wooden cylinder. There are n turns on each meter of length, and each turn of the coil carries current I. The radius of the coil is r. Show that the inductance per unit length of the coil is $3.95r^2n^2$ microhenries per meter. (Note: This is for an infinitely long coil, but is in error by less than 10 per cent if the length of the coil is more than four times its diameter.)

2. A coil is wound on a core as in Example 1, page 99, except that the core is a square frame instead of a ring. Also, the core is iron, not wood. Prove definitely whether or not the magnetic flux can be entirely confined to the iron or will "cut corners" in air.

3. Repeat Example 1, page 99, for a coil wound on a toroidal core of circular cross section. Find the magnetic flux in the core, and the inductance. Check the computed inductance with a value that may be obtained from a handbook.

108   EXAMPLES AND INTERPRETATION

4. Prove that the field of equation 237 has no curl (as is stated on page 106).

5. The idea of *ampere-turns* is a convenient short cut for integrating what vector field over what surface?

6. The idea of *flux-linkages* is a convenient short cut for integrating what vector field over what surface?

7. Using the concepts of this book, what is: (a) voltage, (b) current, (c) inductance, (d) capacitance, (e) electromotive force, (f) magnetomotive force?

8. Outline the important steps in deriving, from the experimental evidence of this book: (a) Kirchhoff's two laws, (b) Lenz's law, (c) Faraday's law of induction, (d) the Biot-Savart law, (e) Ampère's law, (f) Coulomb's law, (g) Poisson's law.

CHAPTER VIII

Maxwell's Hypothesis

The experimental evidence is now before us. From nine experiments described in the preceding chapters, we are fully informed regarding electric and magnetic fields. The results of the experiments are summarized below, and included with them are two assumptions that are, essentially, definitions:

From Experiment I,   an electric field is found to exist and is defined.
From Experiment II,   the electrostatic field is lamellar (without curl).
From Experiment III,   divergence of the electrostatic field is proportional to charge density.
From Experiment IV,   the behavior of dielectric substances is known.
From Experiment V,   Ohm's law is established.
From Experiment VI,   a magnetic field is found to exist and is defined.
From Experiment VII,   a changing magnetic field is found to induce an electric field.
From Experiment VIII,   the magnetostatic field is solenoidal (without divergence).
From Experiment IX,   the curl of the magnetostatic field is proportional to current density.

It is assumed that the dynamic electric field has divergence proportional to charge density. (This is proved in Experiment III for the *static* electric field only.)

It is assumed that the dynamic magnetic field has no divergence. (This is proved in Experiment VIII for the *static* magnetic field only.)

The information obtained from these experiments is the basis of the following discussion of electromagnetism. Either these experiments or others that yield equivalent data must be the foundation of any development of electromagnetic theory.

Information regarding electric and magnetic fields was available to scientists about the middle of the nineteenth century. It was not in the mathematical form in which it is given here; indeed the very concept of an electric or magnetic field was at that time a new idea of Faraday's, considered with doubt by most scientists. The important controversy between believers in "action at a distance" and converts to the newly proposed "field theory" was at its height. Of course, electrostatic and magnetostatic fields had been computed and graphically indicated for the better part of a century, but Faraday was the first (according to James Clerk Maxwell) actually to believe in the existence of the electromagnetic field. Previously, fields had been looked upon as convenient means of visualizing the arrangement of forces that resulted from electric and magnetic action, but to Faraday (as to us) the magnetic field was the actual means by which magnetic force was exerted.

From the time of Ampère's work (1820 to 1825), it had been considered that one wire carrying current exerted a force on another wire carrying current, and no intermediate agency for exerting that force was taken into account. This was the action-at-a-distance theory, and it followed logically Newton's famous law of gravitation, then a century old and universally accepted. Newton's law assumed action at a distance, for it did not consider any medium necessary for the transmission of gravitational force. It was only natural that electrical scientists of the early nineteenth century would follow this illustrious precedent.

Faraday, however, conceived the physical reality of electric and magnetic fields, and Maxwell undertook to express the mathematical relationships involved. It was Maxwell who pointed out that a "displacement current" (as in our equation 177) would simplify and improve the mathematical system. Then Maxwell made a most remarkable proposal as follows: It is known by experiment that *conduction* current produces a magnetic field; *total* current is for mathematical purposes best expressed as the sum of conduction current and displacement current; is it not, then, likely that *displacement* current also produces a magnetic field? Experimental technique did not permit this to be either proved or disproved by direct investigation in Maxwell's time, for the quantities involved were too small. But this hypothesis led to a conclusion of fundamental importance, for Maxwell showed that, if it were true, energy would be transmitted as electromagnetic waves.

The action-at-a-distance theory assumed that electrical action appeared instantaneously at all points, however remote. Maxwell's theory, on the other hand, required that energy be transmitted by waves traveling at a finite speed. This speed of wave propagation

MAXWELL'S HYPOTHESIS

could be computed. Perhaps some experimental verification of the theory would be obtained by studying the velocity of electromagnetic disturbances. Before considering experimental evidence we will follow Maxwell's reasoning that leads to electromagnetic waves.

Maxwell's hypothesis was that, in general, when there are varying electric fields, a magnetic field is produced by the sum of the conduction current and the displacement current. In equation 203, which is

$$\nabla \times \mathbf{H} = \iota \qquad [203]$$

the right-hand side of the equation, according to Maxwell, should be the total current density, as in equation 177, and

$$\nabla \times \mathbf{H} = \iota + \frac{\partial \mathbf{D}}{\partial t} \qquad [240]$$

This is one of the equations known as **Maxwell's equations.** The other is equation 193 of Chapter VI:

$$\nabla \times \mathbf{E} = -\frac{\partial \mathbf{B}}{\partial t} \qquad [193]$$

The other two fundamental field equations are

$$\nabla \cdot \mathbf{B} = 0 \qquad [196]$$

$$\nabla \cdot \mathbf{D} = \rho \qquad [115]$$

These are the basic equations of electromagnetic theory. They are repeated for ready reference in Table III (inside back cover).

It will be noted that Maxwell's equations become beautifully simple and symmetrical when applied in a homogeneous medium in which there is no charge and no conductivity:

$$\nabla \times \mathbf{H} = \epsilon \frac{\partial \mathbf{E}}{\partial t} \qquad [241]$$

$$\nabla \times \mathbf{E} = -\mu \frac{\partial \mathbf{H}}{\partial t} \qquad [242]$$

$$\nabla \cdot \mathbf{H} = 0 \qquad [243]$$

$$\nabla \cdot \mathbf{E} = 0 \qquad [244]$$

It will be well to consider once more the physical meaning of these equations. Equation 241 says that a changing electric field will produce a magnetic field, and equation 242 says that a changing magnetic field will produce an electric field. The latter relation is the familiar principle on which transformers work; the former is Maxwell's hypothesis which says that displacement current as well as conduction current is able to produce a magnetic field. Equation 241 does not contain a term to account for magnetic field produced by actual conduction current because the equation is a simplified one applying to a region in which $\gamma = 0$ and hence $\iota = 0$; equation 240 is the complete equation.

But equations 241 and 242 are particularly interesting to consider relative to the propagation of electric waves. It will be seen at once that, if a changing electric field produces a changing magnetic field, and that in turn produces an electric field which produces a magnetic field, and so on, some kind of a series of energy transfers is started whenever any electric or magnetic disturbance takes place. Energy will be transferred from the electric field to the magnetic, and back to the electric, and so on indefinitely. If (as is actually true) the magnetic energy is not confined to precisely the same location in space as the electric energy from which it is derived, but extends a little beyond, and, if the electric energy derived from that magnetic energy is again a little farther advanced in space, and so on, so that as the energy is changing from form to form it is also being propagated through space, the result may quite reasonably be expected to be a traveling wave of electromagnetic energy.

Consider a somewhat analogous situation. By some means a small volume of water in the middle of a lake is artificially set into vertical oscillatory motion. Perhaps a bucketful of water is suddenly dumped into the lake. Whatever the character of the disturbance, the surface of the water at that point rises and falls in an oscillatory manner. But it is not possible for the bucketful of water to oscillate independently of the water surrounding it. Its periodic excesses and deficiencies of pressure are transmitted to the surrounding water, which thereby receives energy and is, in turn, put into motion. In its resulting undulation it also transfers energy to the next outer region. By this process a wave is propagated across the surface of the lake.

The fundamental reason for the existence of a water wave is this: the motion and pressure of a given volume of water are not independent of the motion and pressure of the water surrounding that volume, and, as the given volume of water is disturbed, it transmits energy to the water next to it.

The fundamental reason for the existence of an electric wave is very similar. A changing magnetic field induces an electric field, both in the region in which the magnetic field is changing and also in the surrounding region; likewise a changing electric field produces a magnetic field in the region in which the change takes place and also in the surrounding region.[1] Consequently when there is a disturbance of either the electric or magnetic conditions in a given region of space, the disturbance cannot be confined to that space. The changing fields within that region will induce fields in the surrounding region also, and those, in turn, in the next surrounding space, and energy is propagated outward. As this action continues a wave of electromagnetic energy is transmitted.

When there is an excess of electromagnetic energy in unbounded space, it cannot stand still, any more than a mound of water can be stationary on the surface of a lake. It cannot merely subside. It can only travel as a wave until the energy is dissipated.

To show that this is indeed the action prescribed by Maxwell's equations, we will develop from them the so-called **wave equations**. We have two equations to begin with, each of which contains both **E** and **H**, and the first step is to solve the equations simultaneously in order to eliminate one of the variables and retain the other. Let us eliminate **H** and thereby obtain an equation in which the only variables are **E** and time. Such an equation will be more easily interpreted.

Before going farther it is well to remark that Maxwell's equations are partial differential equations. Equation 242, for example, equates the rate of change of electric field through space (the curl) to the rate of change of magnetic field with time. It is too much, therefore, to hope for any single simple solution, for that is not commonly to be obtained from simultaneous partial differential equations. As in all problems involving such equations, boundary conditions are all-important.

It was stated above that, according to equation 241, a changing electric field will produce a magnetic field. Strictly, the equation says that a changing electric field will produce a space derivative (curl) of a magnetic field. But it is obvious that, if a magnetic field has some value of curl and therefore varies from point to point in space, it cannot everywhere be zero. Since curl of **H** cannot exist without **H** also existing, we may safely say that a changing electric field produces a magnetic field.

Now we are ready to proceed with the simultaneous solution of equations 241 and 242. First take the curl of each side of equation 242, giving

$$\nabla \times (\nabla \times \mathbf{E}) = -\mu \nabla \times \frac{\partial \mathbf{H}}{\partial t} \qquad [245]$$

[1] Perhaps it is helpful to think of it this way: current flowing in a wire produces magnetic field in the wire, but it also produces magnetic field in space around the wire—space in which no current is flowing. Similarly, displacement current (which results from a changing electric field) produces magnetic field in the region in which displacement current exists (where the electric field is changing) and also in the surrounding region.

On the right-hand side of this equation it may be noted that the curl, which is a partial derivative with respect to distance, operates on a partial derivative with respect to time. It is a well-known mathematical principle that the order of partial differentiation makes no difference, and therefore

$$\nabla \times (\nabla \times \mathbf{E}) = -\mu \frac{\partial}{\partial t} (\nabla \times \mathbf{H}) \qquad [246]$$

(If this is not clear, expand the curl in rectangular coordinates.)

But the curl of **H** is known from equation 241 and may be substituted into equation 246, giving

$$\nabla \times (\nabla \times \mathbf{E}) = -\mu \frac{\partial}{\partial t}\left(\epsilon \frac{\partial \mathbf{E}}{\partial t}\right) = -\mu\epsilon \frac{\partial^2 \mathbf{E}}{\partial t^2} \qquad [247]$$

This equation is entirely in **E**, as desired, but it may be simplified.

The left-hand member of equation 247 is the curl of the curl of a vector. It may be shown as a general mathematical relation that the curl of the curl of any vector is equal to the gradient of the divergence of that vector minus the Laplacian.[2] Symbolically,

$$\nabla \times (\nabla \times \mathbf{A}) = \nabla(\nabla \cdot \mathbf{A}) - \nabla^2 \mathbf{A} \qquad [248]$$

This theorem is applied to equation 247. The divergence of the electric field is zero because it has been assumed that there is no free charge present. The first term of the expansion drops out, leaving simply

$$\nabla^2 \mathbf{E} = \mu\epsilon \frac{\partial^2 \mathbf{E}}{\partial t^2} \qquad [249]$$

To some readers equation 249 will be a familiar form, recognizable as a wave equation. To others its nature will be clearer if the vector quantities are expanded in Cartesian form:

$$\frac{\partial^2 E_x}{\partial x^2} + \frac{\partial^2 E_x}{\partial y^2} + \frac{\partial^2 E_x}{\partial z^2} = \mu\epsilon \frac{\partial^2 E_x}{\partial t^2}$$

$$\frac{\partial^2 E_y}{\partial x^2} + \frac{\partial^2 E_y}{\partial y^2} + \frac{\partial^2 E_y}{\partial z^2} = \mu\epsilon \frac{\partial^2 E_y}{\partial t^2} \qquad [250]$$

$$\frac{\partial^2 E_z}{\partial x^2} + \frac{\partial^2 E_z}{\partial y^2} + \frac{\partial^2 E_z}{\partial z^2} = \mu\epsilon \frac{\partial^2 E_z}{\partial t^2}$$

[2] See footnote 5, page 32.

If a simple special case is now considered in which E_x and E_z do not exist and with E_y a function of x but not of y or z, the equations 250 reduce to

$$E_x = 0 \qquad E_z = 0$$

$$\frac{\partial^2 E_y}{\partial x^2} = \mu\epsilon \frac{\partial^2 E_y}{\partial t^2} \qquad [251]$$

Each of the restrictions leading to equation 251 has a physical meaning that will be discussed in a later chapter.

Equation 251 is the simplest form of the traveling-wave differential equation. Its solution is

$$E_y = f(x - vt) \qquad [252]$$

in which $v = 1/\sqrt{\mu\epsilon}$. The function $f(x - vt)$ represents any function of the quantity $(x - vt)$. [Equation 251 has other solutions also, one of which is $f_2(x + vt)$. We are not at present interested in these other solutions.] To prove that equation 252 is a solution of equation 251, it may be substituted into that equation and the indicated differentiations performed, as in Problem 5.

It is now necessary to recognize that equation 252 describes a transverse wave of constant size and shape that is traveling in the positive direction along the X axis with velocity v. Readers to whom this is unfamiliar will find it helpful to consider specific functions of $(x - vt)$ and to plot $f(x)$ for a number of successive values of time.

Let us consider one such example. First, assume that the wave is traveling in free space; the velocity is then called c (a specific value of v that applies to free space) and, from the definition following equation 252,

$$c = \frac{1}{\sqrt{\mu_0 \epsilon_0}} \qquad [253]$$

Let us assume the wave is a sinusoidal wave that can be described by the function

$$E_y = \sin(x - ct) \qquad [254]$$

Figure 47 shows one cycle of this wave, plotted as a function of x for several different values of time. In other words, if the wave as it traveled through space were visible, Fig. 47 would be a succession of snapshots of it, taken at the instants at which ct equals 0, $\pi/6$, $\pi/3$, and $\pi/2$. It is apparent from the equation that all these will be sine curves of the same shape and amplitude. But the curve for which $ct = \pi/6$ will be displaced with reference to the one for which $ct = 0$, and each of its corresponding values will occur at a value of x greater by $\pi/6$. That is, to keep the quantity in parentheses unchanged, and therefore to have the same value of E_y, x will have to increase as time increases. The result will be a wave moving from left to right as time passes. Time and distance are related by the factor c as by a velocity. (It

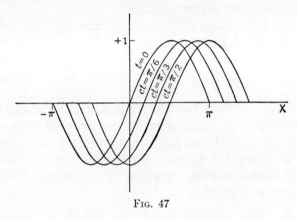

Fig. 47

should nevertheless be noticed that no physical entity is moving at velocity c. The electric field is not moving; it is merely changing in such a way that if it were visible there would appear to be waves of that velocity.)

The wave of equation 252 is *transverse* because the electric field E_y is in the y direction and the wave is propagated in the x direction.

Equation 249 is obtained by eliminating **H** and retaining **E** in equations 241 and 242. If the opposite procedure is followed, eliminating **E** and retaining **H**, a similar wave equation is obtained for the magnetic vector:

$$\nabla^2 \mathbf{H} = \mu\epsilon \frac{\partial^2 \mathbf{H}}{\partial t^2} \qquad [255]$$

The solution of this equation is a traveling wave of **H**.

It is evident that it is not possible to have an electric disturbance without having a magnetic one also, and vice versa, and it will be found that every electromagnetic wave has an electric portion and a magnetic portion traveling along together. To have one without the other would be analogous to having a water wave in which there is motion without displacement, or displacement without motion. Although there are separate wave equations for **E** and **H**, they represent physically inseparable quantities.

Recognition of the fact that his electromagnetic equations had a traveling-wave solution led James Clerk Maxwell to very interesting speculations. In the first place, if it could be shown that electromagnetic waves exist, Maxwell's hypothesis relating to the ability of a displacement current to produce a magnetic field would be justified, for without that hypothesis no wave solution would result. The situation would be like a lake in which the water has weight but no mass—waves could not exist.

If electromagnetic waves do exist they should travel in free space with the velocity c, from equation 253. But c is a known quantity, for it can be computed from μ_0 and ϵ_0. These latter are dimensional constants that can be evaluated by measuring electrostatic and magnetic forces. By balancing one force against the other in an ingenious kind of current balance, Maxwell found, about 1865, that the numerical value of c is a little less than 3×10^8. This being so, electromagnetic waves should travel, if they exist, at the rate of some 3×10^8 meters or 300,000 kilometers per second.

In what medium do electric waves exist? In Maxwell's time it was commonly accepted that ordinary visible light is a wave motion of a luminiferous aether, an aether pervading all space and all material, without weight but with remarkable elastic properties that permit it to propagate transverse waves. Maxwell's wave equations indicated that electromagnetic disturbances were propagated as transverse waves in some similar medium. Could it be that light is merely a form of electric wave?

Faraday had speculated upon this possibility several years earlier and had pointed out that, whereas one infinite and all-pervasive imponderable aether is a severe strain upon one's imagination, belief in two co-existent, infinite, all-pervasive, and imponderable aethers, one for light and one for electricity, is simply beyond the limits of credulity. Therefore he suspected, on this basis alone, that light is an electric phenomenon.

Maxwell had a more substantial reason for coming to the same conclusion, for he collected all the best measurements of the speed of light in vacuum, and he found that the average of those available to him was amazingly close to 300,000 kilometers per second, which was the velocity that he had predicted for electric waves from laboratory measurement of electrical units.

This being so, Maxwell's hypothesis was substantiated, and it appeared at least highly probable that light is indeed an electric wave.

Although Maxwell accepted these conclusions, many other scientists did not until Hertz, about twenty-five years later, proved the physical existence of electromagnetic waves. He accomplished this by showing their interference, reflection, and refraction. After that it was no longer possible to doubt that electromagnetic energy is propagated as wave motion. It is still safest to avoid the embarrassing question of the character of the medium in which such waves are transmitted. The best we can do is to follow Faraday's "shadow of a speculation" and "dismiss the aether but keep the vibrations."

PROBLEMS

1. Expand Maxwell's equations 241 and 242 in rectangular coordinates. By equating like components, derive from each three scalar equations (as in equations 264 to 266).
2. Prove equation 248 by expanding in Cartesian components.
3. Find the Laplacian of H, equation 255. Assume zero conductivity.
4. Extend Problem 3 by assuming the region to be conducting.
5. Prove that equation 252 is a solution of the wave equation 251. The composite function $f(x - vt)$ is differentiated according to methods discussed in calculus books.
6. Show that $E_y = f(x - vt)$ is a solution of equation 251 using the following functions: $f(x) = Ae^{j\beta x}$, $f(x) = A(e^{-ax} - e^{-bx})$, $f(x) = xe^{-ax}$, $f(x) = \sin ax \cos bx$, $f(x) = A_1 \sin \beta x + A_2 \sin 2\beta x$, $f(x) = a_0 + a_1 x + a_2 x^2 + a_3 x^3$. (Note: Which of these functions can you identify as having practical importance as traveling waves?)
7. Show that $E_y = f(x + vt)$ is a solution of equation 251 for one of the functions given in Problem 6 for $f(x)$.
8. Derive equations 243 and 244 from equations 241 and 242, assuming the fields are not static.

CHAPTER IX
Plane Waves

When light starts from a point on the sun it radiates outward in all directions and travels as a spherical wave. Part of that wave eventually reaches the earth, where we can observe it and measure it in our laboratories. The part that reaches us appears as a plane wave. That is simply because we are limited by the size of the laboratory (or at most by the size of the earth) and can observe only a very small part of the whole spherical wave. Just as the ocean appears flat to a man who can see only a few miles around him, so the wave appears plane to an observer who can study only a small part of it.

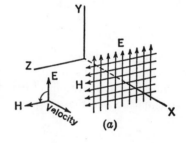

Much the same thing is true of radio waves. From the point of view of a transmitting antenna, the wave is radiated in all directions. Indeed the radiation pattern is an important practical consideration. But, from the point of view of the receiving antenna, any wave from a station several miles away is practically a plane wave. (This state-

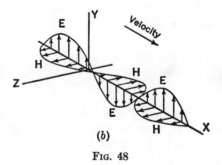

Fig. 48

ment neglects possible complications such as the effect of ground or reflections from the ionosphere.)

Let us consider a simple example of a plane wave. In Fig. 48a a cross section of a plane wave is indicated. Electric and magnetic vectors are shown in a plane parallel to the Y-Z plane. The wave is traveling from left to right along the X axis; it fills all the region shown in the figure, but only one cross section is indicated. The electric field

119

PLANE WAVES

is vertical; it is uniform throughout the plane in direction and magnitude.

Electric Field. We must describe this wave mathematically. First, in common with all electromagnetic waves, it must satisfy the wave equation:

$$\nabla^2 \mathbf{E} = \mu\epsilon \frac{\partial^2 \mathbf{E}}{\partial t^2} \qquad [249]$$

Next, because the electric field is entirely in the y direction,

$$E_x = 0 \qquad E_z = 0 \qquad [256]$$

Finally, it is assumed that the electric field strength is uniform throughout the plane of the wave and does not vary from point to point in that plane as either y or z is changed. Thus

$$\frac{\partial \mathbf{E}}{\partial y} = 0 \qquad \frac{\partial \mathbf{E}}{\partial z} = 0 \qquad [257]$$

Equation 249 can be expanded into three equations representing its Cartesian components, as in equation 250, page 114. When this is done and the simplifying conditions of equations 256 and 257 are introduced, the wave equation becomes merely

$$\frac{\partial^2 E_y}{\partial x^2} = \mu\epsilon \frac{\partial^2 E_y}{\partial t^2} \qquad [258]$$

The solution of this equation [1] will be a mathematical expression of our wave.

The complete wave solution of equation 258 is

$$E_y = f_1(x - vt) + f_2(x + vt) \qquad [259]$$

The first part of this solution was discussed in Chapter VIII and represents a wave traveling in the direction of the X axis, as in Fig. 47. The function f_1 may be any function and depends upon the type of disturbance that starts the wave. Most radio waves are approximately sinusoidal functions, and for these f_1 would be a sine or cosine function of $(x - vt)$.

The second part of the solution is also a traveling wave, but it represents a wave traveling in the *negative* x direction. Thus the complete

---

[1] It may be noted that this is identical with equation 251, which resulted from mathematical assumptions similar to equations 256 and 257 but without any physical interpretation at the time.

solution describes two traveling waves,[2] one going in each direction, simultaneously passing through the same space, and the equation informs us that they travel independently of each other. For the present it will be convenient to let $f_2 = 0$. This leaves only a pure traveling wave.

Magnetic Field. So far only the electric component of the wave has been discussed. This must be accompanied by a magnetic component which can be determined from the electric component by means of Maxwell's equations. The general method is as follows. One of Maxwell's equations (Table III) says

$$\frac{\partial \mathbf{B}}{\partial t} = -\nabla \times \mathbf{E} \qquad [260]$$

$\mathbf{E}$ is known, and its curl can be found. Thus, the time derivative of $\mathbf{B}$ is determined. Integration with respect to time then gives $\mathbf{B}$.

This solution for the magnetic field can proceed to a certain extent in general terms, but it is more intelligible to consider a specific wave. It has been mentioned that the most common wave in practice is approximately sinusoidal. Let us, therefore, assume a wave in which the electric field is described by

$$E_x = 0 \qquad E_y = E_m \cos \beta(x - vt) \qquad E_z = 0 \qquad [261]$$

It will be seen that this is consistent with equation 259, for it gives the y component of the electric field as a function of $(x - vt)$. It is therefore a solution of the wave equation. The coefficients E_m and β can have any value; from the mathematical point of view they are merely arbitrary constants, but physically the former determines the amplitude of the wave and the latter (known as the **phase constant**) determines the frequency of its sinusoidal variation. Such a sinusoidal wave is indicated in Fig. 48b.

A more usual form of the same equation is obtained by letting $\beta v = \omega$ and writing

$$E_x = 0 \qquad E_y = E_m \cos (\omega t - \beta x) \qquad E_z = 0 \qquad [262]$$

From its association with time in this equation, ω is identified as $2\pi f$, f being frequency. Introducing λ as wavelength, and using v as the

---

[2] Other terms might be added to equation 259 as a solution of equation 258. A constant term could be added, and terms containing either x or t in the first degree, for the second derivatives of all such terms would vanish. But such terms do not represent wave action and may be disregarded in the present discussion.

velocity of phase propagation in a given medium, the following simple but frequently useful relations [3] are apparent:

$$v = \frac{\omega}{\beta} = 2\pi\frac{f}{\beta} \qquad \lambda = \frac{v}{f} = \frac{2\pi}{\beta} \qquad [263]$$

The phase constant β is inversely proportional to wavelength; λ is measured in meters per cycle, and β in radians per meter.

The first step in determining the magnetic field is to expand Maxwell's equation 260 into its component parts:

$$\frac{\partial B_x}{\partial t} = -\left(\frac{\partial E_z}{\partial y} - \frac{\partial E_y}{\partial z}\right) \qquad [264]$$

$$\frac{\partial B_y}{\partial t} = -\left(\frac{\partial E_x}{\partial z} - \frac{\partial E_z}{\partial x}\right) \qquad [265]$$

$$\frac{\partial B_z}{\partial t} = -\left(\frac{\partial E_y}{\partial x} - \frac{\partial E_x}{\partial y}\right) \qquad [266]$$

For the particular wave under consideration, as described in equations 262, these components of Maxwell's equation become

$$\frac{\partial B_x}{\partial t} = 0 \qquad \frac{\partial B_y}{\partial t} = 0 \qquad [267]$$

$$\frac{\partial B_z}{\partial t} = -\frac{\partial E_y}{\partial x} = -\beta E_m \sin(\omega t - \beta x) \qquad [268]$$

The components of the magnetic field are found by integration with respect to time:

$$B_x = 0$$
$$B_y = 0 \qquad [269]$$
$$B_z = \frac{\beta}{\omega} E_m \cos(\omega t - \beta x) = \frac{\beta}{\omega} E_y$$

Remembering that ω/β is the velocity of phase propagation,

$$E_y = vB_z \qquad [270]$$

[3] These and others are collected for ready reference in Table III, inside back cover.

Thus, in a simple plane wave traveling in a dielectric medium, the electric and magnetic components of the wave are identical in form and perpendicular in direction. They are both perpendicular to the direction of travel of the wave. In magnitude, the electric field in volts per meter is equal to the magnetic flux density in webers per square meter times the velocity in meters per second. The velocity v was shown to be $1/\sqrt{\mu\epsilon}$; when the wave is in free space, this has the numerical value of c, about 2.998×10^8—or, for most purposes, 3×10^8 meters per second.

Since $\mathbf{B} = \mu\mathbf{H}$, we can also write

$$E_y = v\mu H_z = \frac{\mu}{\sqrt{\mu\epsilon}} H_z = \sqrt{\frac{\mu}{\epsilon}} H_z \qquad [271]$$

This gives a relation between E and H. The ratio of E to H in a wave is so often useful that it is given a special symbol η, and a name, **intrinsic impedance**. Equation 271 shows that, for a simple plane wave in a dielectric medium,

$$\eta = \sqrt{\frac{\mu}{\epsilon}} \qquad [272]$$

Intrinsic impedance is somewhat analogous to the characteristic impedance of a transmission line. It has the dimensions of ohms, for E is volts per meter and H is amperes per meter. The value of η in free space is $\sqrt{\mu_0/\epsilon_0} = 4\pi c \cdot 10^{-7}$; this is very nearly 376.7, and for most purposes it is remembered as 377 or 120π.

Lines of electric and magnetic field are shown in Fig. 48a. Throughout the plane indicated in that figure, **E** and **H** are uniform. If the plane shown in the figure is visualized as being fixed in space, **E** and **H** in that plane are constantly changing with time. If, on the other hand, the plane indicated is visualized as advancing along the X axis with the speed of light, **E** and **H** in that plane are constant and unchanging. This is, indeed, the distinctive and defining quality of a *plane wave*.

Figure 48b is a graphical representation of the sinusoidal wave. It is traveling in the positive direction along the X axis. Vectors of **E** and **H** are shown, and each arrow represents the electric or magnetic intensity throughout the entire plane in which it lies. The length of each vector shows the strength of the field. The fields vary sinusoidally along the X axis, and, when it is remembered that the entire wave train is moving along the X axis, it is apparent that at any fixed point the electric and magnetic intensities vary sinusoidally with time.

Neither Fig. 48a nor 48b is a complete representation of the wave. To show the wave as it exists in three-dimensional space would require a combination of these diagrams, with the distribution of the electric and magnetic fields shown throughout many planes. Even this would be only an instantaneous picture of the wave and would fail to show its motion; for a complete picture the imagination must be called upon to visualize what cannot be drawn upon paper.

Polarization. This wave is a plane wave. It is also a polarized wave. If the electric vector oscillates as the wave passes but maintains the same direction, the wave is said to be polarized. Mathematically, equation 256 specifies that the wave is polarized, and equations 256 and 257 together specify a polarized plane wave. In the general case of an unpolarized wave, the electric and magnetic vectors change direction as well as magnitude as the wave passes. (The illustration of transverse waves traveling along a stretched rope is generally familiar: if the rope vibrates in a single plane its wave motion is polarized.)

The polarization of radio waves is determined mainly by the transmitting antenna. The received "ground wave" from a vertical antenna, as commonly used in the standard broadcast frequency band, is vertically polarized because the electric field from a transmitting antenna to ground is substantially vertical. High-frequency antennas, on the contrary, are usually horizontal, and the electric field from end to end of the antenna produces a horizontally polarized wave at the receiver. Ordinary light is not polarized; but, if it passes through some material that reflects, refracts, or absorbs one component of vibration while allowing the other to pass, the portion of light that is transmitted is then polarized.[4]

Exponential Notation. In the preceding sections the electric field of a wave was described by equation 261, using a cosine function to represent the sinusoidal variation of the field as a function of x and also as a function of time:

$$E_y = E_m \cos(\omega t - \beta x) \qquad [261]$$

Another possible mathematical formulation is to use an exponential function and write

$$E_y = \text{Real part of } E_m\, e^{j(\omega t - \beta x)} \qquad [273]$$

The equivalence of these two expressions appears when the exponential

---

[4] In optics the plane of polarization is the plane (parallel to the direction of propagation) that contains the *magnetic* vector. This definition is arbitrary and was adopted before the electromagnetic nature of light was suspected. In radio work it is common to consider the wave polarized in the direction of the *electric* vector.

is expanded according to the general formula [5]

$$e^{jx} = \cos x + j \sin x$$

giving

$$E_y = \text{Real part of } \{E_m[\cos(\omega t - \beta x) + j \sin(\omega t - \beta x)]\}$$
$$= E_m \cos(\omega t - \beta x) \qquad [274]$$

It must be emphasized that the physical quantity (E_y in these equations) is represented by only the real part of the exponential function, and the equations should contain the words "Real part of" or some symbol with the same meaning. Nevertheless, it is customary to omit any such symbol and write

$$E_y = E_m\, e^{j(\omega t - \beta x)} \qquad [275]$$

leaving the reader to understand that only the real part of the exponential function is to be used. This practice may occasionally lead to trouble, as will be seen later, but difficulties can always be removed by writing "Real part of" where it belongs.

Use of the exponential function is very helpful in manipulation of equations, and it is common in alternating-current circuit work as well as field theory. Indeed, use of the exponential function is tacitly assumed whenever "impedance" is mentioned, for impedance is the ratio of an exponential voltage to an exponential current—not the ratio of voltage to current when expressed as a sine or cosine function.

Propagation in Conducting Material. Until now we have considered waves traveling in a perfect dielectric medium. The restriction to a perfect dielectric was introduced by setting $\gamma = 0$ in Maxwell's equations. The wave solution can be extended to include waves in material with finite conductivity, but the mathematical difficulty becomes very great unless the solution is restricted to sinusoidal waves. However, a solution for sinusoidal waves is what is usually wanted, especially since any periodic wave can be analyzed into Fourier components. Therefore, in the following work we shall assume from the beginning that the electric and magnetic fields vary sinusoidally with time. Having assumed this, we know that both **E** and **H** will contain a time-variation factor that may be written $e^{j\omega t}$, and we write

$$\mathbf{E} = \mathbf{E}_0\, e^{j\omega t} \qquad \mathbf{H} = \mathbf{H}_0\, e^{j\omega t} \qquad [276]$$

$\mathbf{E}_0$ and $\mathbf{H}_0$ are vector fields, functions of space coordinates but not of time. They may be considered snapshots of the fields taken at the

---

[5] The italic j is by definition $\sqrt{-1}$, and it is not the same as the bold-faced **j** which is one of the unit vectors of a Cartesian coordinate system. Vector notation is not to be confused with complex-quantity notation. See also footnote 1 on page 13.

126   PLANE WAVES

instant $t = 0$. We now solve for $\mathbf{E}_0$, and the first step is to write the wave equation in a form to include conductivity, but simplified by being restricted to sinusoidal waves. We start with the general equations (from 193 and 240):

$$\nabla \times \mathbf{E} = -\mu \frac{\partial \mathbf{H}}{\partial t} \qquad [277]$$

$$\nabla \times \mathbf{H} = \gamma \mathbf{E} + \epsilon \frac{\partial \mathbf{E}}{\partial t} \qquad [278]$$

Substituting equations 276 and differentiating with respect to time,

$$\nabla \times \mathbf{E}_0 e^{j\omega t} = -j\omega\mu \mathbf{H}_0 e^{j\omega t} \qquad [279]$$

$$\nabla \times \mathbf{H}_0 e^{j\omega t} = \gamma \mathbf{E}_0 e^{j\omega t} + j\omega\epsilon \mathbf{E}_0 e^{j\omega t} \qquad [280]$$

Since curl is differentiation with respect to distance, the exponential time factor can be divided out, leaving

$$\nabla \times \mathbf{E}_0 = -j\omega\mu \mathbf{H}_0 \qquad [281]$$

$$\nabla \times \mathbf{H}_0 = (\gamma + j\omega\epsilon)\mathbf{E}_0 \qquad [282]$$

These are purely space equations; time has been eliminated.

Taking the curl of each side of equation 281, substituting into the result equation 282 to eliminate $\mathbf{H}_0$, and using the identity of equation 248 (with the assumption that there is no free charge present),

$$\nabla \times \nabla \times \mathbf{E}_0 = -j\omega\mu(\nabla \times \mathbf{H}_0)$$
$$= -\nabla^2 \mathbf{E}_0 = -j\omega\mu(\gamma + j\omega\epsilon)\mathbf{E}_0 \qquad [283]$$

For convenience, let us use Γ^2 to represent the coefficient:

$$\Gamma^2 = j\omega\mu(\gamma + j\omega\epsilon) \qquad [284]$$

so that

$$\nabla^2 \mathbf{E}_0 = \Gamma^2 \mathbf{E}_0 \qquad [285]$$

This is the wave equation for a sinusoidal wave, including conductivity.

Now let it be assumed, as it was on pages 115 and 120, that we are interested in a plane wave polarized in the Y direction, so that

$$E_x = E_z = 0 \qquad \frac{\partial E_y}{\partial y} = \frac{\partial E_y}{\partial z} = 0 \qquad [286]$$

Then equation 285 becomes merely

$$\frac{\partial^2 E_{0y}}{\partial x^2} = \Gamma^2 E_{0y} \qquad [287]$$

PROPAGATION IN CONDUCTING MATERIAL

and the solution of the wave equation for a polarized plane wave is

$$E_{0y} = E_m\, e^{\pm \Gamma x} \qquad [288]$$

That this solution is correct is obvious when it is substituted into equation 287, and E_m can be any constant. E_y, the only component of the electric field, is therefore

$$E_y = E_{0y}\, e^{j\omega t} = E_m\, e^{j\omega t}\, e^{\pm \Gamma x}$$
$$= E_m\, e^{j\omega t \pm \Gamma x} \qquad [289]$$

The meaning of the symbols may be reviewed. **E** is the electric field; $\mathbf{E}_0$ is its initial value and is not a function of t; E_{0y} is the Y component of $\mathbf{E}_0$. E_m is a constant (it is the value of E_{0y} at the origin) and is not a function of x, y, z, or t.

If equation 289 is applied to propagation in a perfect dielectric, in which conductivity is zero, the equation reduces to a familiar form. With $\gamma = 0$ in equation 284,

$$\Gamma = j\omega \sqrt{\mu\epsilon} = \frac{j\omega}{v} = j\beta$$

Substituting $j\beta$ for Γ in equation 289 (retaining only the minus sign) gives

$$E_y = E_m\, e^{j(\omega t - \beta x)} \qquad [290]$$

Considering the meaning of the exponential notation, this is identical [6] with equation 262. (The $\pm$ sign in equation 289 means that the wave we are discussing may, within the restrictions laid down, be traveling in either direction, as discussed on page 120.)

In general, from equation 284

$$\Gamma = j\omega \sqrt{\mu\epsilon \left(1 + \frac{\gamma}{j\omega\epsilon}\right)} \qquad [294]$$

[6] In equation 262, lacking the convenience of exponential formulation, E_m is necessarily real, and phase is indicated by an additional term in the argument of the cosine. Thus, if a phase difference of ϕ is to be included, equation 262 becomes

$$E_y = E_m \cos(\omega t - \beta x + \phi) \qquad [291]$$

The corresponding form of 289 is

$$E_y = |E_m|\, e^{j(\omega t - \beta x + \phi)} \qquad [292]$$

which may equally well be written

$$E_y = |E_m|\, e^{j\phi}\, e^{j(\omega t - \beta x)} = E_m\, e^{j(\omega t - \beta x)} \qquad [293]$$

where $E_m = |E_m|\, e^{j\phi}$. This inclusion of phase in E_m is customary.

If conductivity is zero, Γ simply reduces to $j\omega\sqrt{\mu\epsilon} = j\beta$, as noted above, but, if conductivity is not zero, Γ is complex. Γ will have a real part and an imaginary part, and, if these are called α and $j\beta$, respectively, we may write

$$\Gamma = \alpha + j\beta \qquad [295]$$

The values of α and β are computed from equation 294. Using these values in equation 289, the traveling wave in conducting material is described by

$$E_y = E_m e^{-\alpha x} e^{j(\omega t - \beta x)} \qquad [296]$$

(This describes a wave traveling in the positive x direction, for only the negative sign preceding Γ in equation 289 has been retained.)

The wave in conducting material is similar to the wave in space or in a perfect dielectric except that it grows smaller as a result of losing energy as it travels. The factor $e^{-\alpha x}$ in equation 296 shows this attenuation. If the conductivity of a dielectric is small, the wave travels with little attenuation, for α is small. In a good conductor, such as metal, the attenuation is so rapid that practically all energy carried by a wave into the metal is lost in a small fraction of a millimeter (see problem 9, page 136). In a moderate conductor, such as moist earth, radio waves of broadcast frequency can penetrate several meters before being almost entirely dissipated. Obviously it is only the energy that enters the conducting material that is attenuated in this way; the reflection of energy from a conducting surface is a different problem that will be considered in Chapter X.

The phase constant β in conducting material is different from the phase constant in non-conducting material of the same dielectric constant and permeability; β increases with conductivity (see problem 11, page 136). Hence wavelength corresponding to a given frequency becomes smaller, and velocity of propagation [7] is less. In a substance that has the conductivity of a poor insulator (a quasi-dielectric), the change in wavelength resulting from conductivity may be inappreciable, but, in a fairly good conductor, the wavelength is a small fraction of the wavelength in free space.

The magnetic field of a traveling wave in conducting material is not in phase with the electric field, as we found it to be in a perfect dielec-

---

[7] Velocity of propagation $v = j\omega/j\beta$. Some authors call $j\omega/\Gamma$ the "complex velocity" and use it to account for attenuation as well as speed. In a somewhat similar manner, $\epsilon\left(1 + \dfrac{\gamma}{j\omega\epsilon}\right)$ may be called the "complex dielectric constant." $Q = \omega\epsilon/\gamma$ is a convenient factor for some purposes, partially analogous to the Q of a resonant circuit. These concepts, confusing at first, become useful with experience.

tric. This appears when **H** is found from **E** by means of equation 281:

$$\nabla \times \mathbf{E}_0 = -j\omega\mu\mathbf{H}_0 \qquad [281]$$

In the plane wave under discussion

$$\nabla \times \mathbf{E}_0 = \mathbf{k}\frac{\partial E_{0y}}{\partial x}$$

$$= \mathbf{k}\frac{\partial}{\partial x} E_m e^{-\Gamma x}$$

$$= -\mathbf{k}\Gamma E_m e^{-\Gamma x} \qquad [297]$$

Introducing this value for curl into equation 281:

$$j\omega\mu\mathbf{H}_0 = \mathbf{k}\Gamma E_m e^{-\Gamma x} \qquad [298]$$

or

$$H_z = \frac{\Gamma}{j\omega\mu} E_y \qquad [299]$$

from which the intrinsic impedance is

$$\eta = \frac{E_y}{H_z} = \frac{j\omega\mu}{\Gamma} \qquad [300]$$

Using equation 284 to expand Γ,

$$\eta = \sqrt{\frac{\mu}{\epsilon\left(1 + \dfrac{\gamma}{j\omega\epsilon}\right)}} \qquad [301]$$

This expression for intrinsic impedance clearly reduces to $\sqrt{\mu/\epsilon}$, as in a perfect dielectric (see equation 272), if the conductivity γ is zero. But in a conducting medium it is a complex quantity and, since $E_y = \eta H_z$, it describes a phase difference as well as a difference in magnitude between electric and magnetic fields of a traveling wave in conducting material. Since η is a complex quantity in the first quadrant, the electric field leads the magnetic field in time phase.

Dielectric Loss. In good insulating materials, of the kind ordinarily used for their dielectric qualities, loss owing to actual conductivity of the material is negligible. Another source of loss, however, known as **dielectric hysteresis,** is of the greatest importance.

In an alternating electric field, energy is stored and released during each half cycle. If there is some material substance present in the electric field, the recovery of energy is not complete; a small fraction of the stored energy is lost, possibly owing to a molecular frictional effect. This loss is repeated each half cycle, and hence the power loss is proportional (or approximately proportional) to frequency.

Because dielectric hysteresis loss is nearly proportional to frequency, it is convenient to express the loss in a given substance in terms of a *power factor*. This is the power factor of an otherwise perfect condenser in which the given substance is used as the dielectric material. Power factor is independent of size and shape of the dielectric material, but it usually increases with temperature and humidity; it is reasonably independent of frequency, but, since it may change appreciably if the order of magnitude of the frequency is changed, it is safest to determine the power factor for the approximate frequency to which it is to be applied. Dielectric power factors range from the order of 0.0005 for mica, quartz, and polystyrene, to 0.005 for glass and steatite, and to 0.05 for phenol products.[8]

It is customary to account for dielectric loss by determining an equivalent conductivity that would cause the same amount of loss at the given frequency, and to use this value in the equations of the preceding section. The equivalent conductivity is found by considering 1 square meter of a condenser in which the dielectric material is 1 meter thick; if loss is conductive, the power factor is $\gamma/\omega\epsilon$ (this neglects the difference between the sine and tangent of the small power factor angle). The equivalent conductivity of a dielectric to represent hysteresis loss of known power factor is therefore $\omega\epsilon \times$ (power factor).

Knowing the power factor of a medium, the attenuation of the field strength of a traveling wave is readily expressed in terms of the power factor. Using the first term of a binomial-series expansion of Γ, a good approximation is

$$\alpha = \tfrac{1}{2}\omega\sqrt{\mu\epsilon} \times \text{(power factor)}$$

Converting to *power* loss in decibels (multiplying α by 8.686), gives loss $= 9.1 f \sqrt{\kappa} \times$ (power factor) $\times 10^{-8}$ decibels per meter, f being frequency in cycles per second and κ the relative dielectric constant of the medium,[9] with $\mu = \mu_0$.

[8] Values are given in radio handbooks. See, for instance, *Radio Engineers' Handbook*, F. E. Terman, McGraw-Hill Book Co., 1943, page 111; *Microwave Transmission Design Data*, Sperry Gyroscope Co., 1944.

[9] Although derived for a specific form of wave, this formula is fairly general. It gives attenuation owing to dielectric hysteresis in a coaxial transmission line with solid insulation—or, indeed, for any line with a *TEM* wave. If divided by the guide factor G of Table VIII, page 212, it gives the attenuation owing to dielectric loss in a dielectric-filled wave guide.

Power and the Poynting Vector. A very important aspect of wave propagation is the flow of power through space. It is apparent that a traveling wave carries energy with it, as a radio wave, for example, carries energy from the transmitter to the receiver.

As a wave passes through an imaginary surface in space, its energy will pass through that surface, and at any instant there will be a flow of power through each unit of area of the surface. This quantity, in watts per square meter, will be denoted by the symbol **P**. The product **P** · **a** is power passing (at a given instant) through an area **a**. **P** is a vector quantity, called the Poynting vector after a mathematician of the nineteenth century, and, when flux lines of the vector field of **P** are drawn, they show the flow of electromagnetic energy. The Poynting vector field is remarkably useful in electrodynamics, and the mathematical formulation of the Poynting vector is much simpler than might be expected.

Consider a region of space, enclosed within an imaginary surface. The rate at which electromagnetic energy flows out of this region is found by integrating **P** over the enclosing surface. Thus

$$\text{Outward flow of power} = \oint \mathbf{P} \cdot d\mathbf{a} \qquad [302]$$

But, if energy is flowing out of the region, there must be a corresponding loss of electromagnetic energy stored within the region. Electromagnetic energy is the sum of the electric and magnetic energies given by the volume [10] integrals:

$$\text{Electric energy} = \tfrac{1}{2} \int \mathbf{D} \cdot \mathbf{E} \, d\mathcal{U} \qquad [157]$$

$$\text{Magnetic energy} = \tfrac{1}{2} \int \mathbf{B} \cdot \mathbf{H} \, d\mathcal{U} \qquad [221]$$

$$\text{Total energy} = \tfrac{1}{2} \int (\mathbf{B} \cdot \mathbf{H} + \mathbf{D} \cdot \mathbf{E}) \, d\mathcal{U} \qquad [303]$$

The rate at which this stored energy diminishes is found by differentiation:

$$\left. \begin{array}{l} \text{Rate of decrease} \\ \text{of stored energy} \end{array} \right\} = -\frac{\partial}{\partial t} \frac{1}{2} \int (\mathbf{B} \cdot \mathbf{H} + \mathbf{D} \cdot \mathbf{E}) \, d\mathcal{U} \qquad [304]$$

[10] In this chapter "$\mathcal{U}$" is used as the symbol for volume, to avoid confusion with the use of v for velocity.

Assuming that electrical energy is not being changed to heat by flow of current within the region, an assumption that is correct if there is zero conductivity within the region (as, for example, in free space, or in any perfect insulator), there can be a decrease of stored energy only if there is an equal outward flow of power. Equating 302 and 304:

$$\oint \mathbf{P} \cdot d\mathbf{a} = -\frac{\partial}{\partial t}\frac{1}{2}\int (\mathbf{B} \cdot \mathbf{H} + \mathbf{D} \cdot \mathbf{E})\, d\mathcal{V}$$

$$= -\frac{1}{2}\int \frac{\partial}{\partial t}(\mu \mathbf{H} \cdot \mathbf{H} + \epsilon \mathbf{E} \cdot \mathbf{E})\, d\mathcal{V} \qquad [305]$$

When the indicated differentiation is performed, the right-hand member becomes

$$-\int \left(\mu \mathbf{H} \cdot \frac{\partial \mathbf{H}}{\partial t} + \epsilon \mathbf{E} \cdot \frac{\partial \mathbf{E}}{\partial t}\right) d\mathcal{V} = -\int \left(\mathbf{H} \cdot \frac{\partial \mathbf{B}}{\partial t} + \mathbf{E} \cdot \frac{\partial \mathbf{D}}{\partial t}\right) d\mathcal{V} \qquad [306]$$

Now Maxwell's equations are used to substitute for the time derivatives, giving

$$+\int [\mathbf{H} \cdot (\nabla \times \mathbf{E}) - \mathbf{E} \cdot (\nabla \times \mathbf{H})]\, d\mathcal{V} \qquad [307]$$

As a purely mathematical theorem, applying to any vector field, it can be shown that

$$\nabla \cdot (\mathbf{M} \times \mathbf{N}) = \mathbf{N} \cdot (\nabla \times \mathbf{M}) - \mathbf{M} \cdot (\nabla \times \mathbf{N}) \qquad [308]$$

The right-hand member of this equation corresponds exactly with the quantity in brackets in expression 307, and substitution into 307 gives

$$\int \nabla \cdot (\mathbf{E} \times \mathbf{H})\, d\mathcal{V} \qquad [309]$$

Divergence is here integrated through a volume, and by Gauss's theorem we may substitute for this an integration over the surface enclosing that volume. When this is done, and the result is substituted for the right-hand member of equation 305,

$$\oint \mathbf{P} \cdot d\mathbf{a} = \oint (\mathbf{E} \times \mathbf{H}) \cdot d\mathbf{a} \qquad [310]$$

Both sides of equation 310 are surface integrals, and both are integrated over the same surface enclosing an arbitrary region of space.

POWER AND THE POYNTING VECTOR

Equation 310 is clearly satisfied if the Poynting vector is

$$\mathbf{P} = \mathbf{E} \times \mathbf{H} \qquad [311]$$

and thus the flow of power [11] in wave motion is obtained.

This derivation of the Poynting vector considers a region without conductivity, thereby eliminating resistance loss of energy. This is for simplicity only, and if conductivity is taken into account the result is

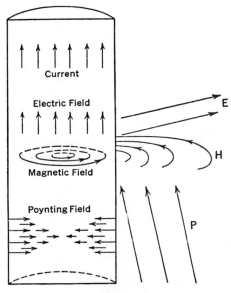

Fig. 49

exactly the same: equation 311 expresses the flow of electromagnetic energy in either a conducting or non-conducting region.

As a simple example of the Poynting vector field, consider a long cylindrical conductor carrying current. In Fig. 49 a steady current is flowing upward in a cylindrical conductor, the front half of the conductor being cut away in the diagram. The electric field within the conductor is correspondingly uniform and upward. The electric field outside of the conductor is much stronger, having a tangential component equal to the

---

[11] Equation 310 is equally well satisfied if there is added to 311 a function for which the integral over every closed surface is zero; a function, that is, with zero divergence. Hence, although the integral of **P** over a surface may commonly be taken to represent power flow, this idea may sometimes lead to absurd interpretations (i.e., if an electrostatic field and the field of a permanent magnet exist in the same region). But the integral of **P** over a *closed* surface is always the true outward flow of power.

field within the conductor and a radial component that terminates on some other part of the circuit. The magnetic field within the conductor is circular, and its strength is proportional to the radius. The Poynting field within the conductor, being **E** × **H**, is radially inward, growing weaker as it penetrates the conductor.

The increasing weakness of the Poynting field indicates the consumption of energy. Energy enters the surface of the conductor and flows toward the center; it is used to supply resistance loss in the conductor, and, as the center of the conductor is approached, the inward flow of energy decreases to zero. This energy is supplied from the external electromagnetic field. The Poynting field outside of the conductor is primarily parallel to the conductor, showing that energy is being carried in the direction of the conductor (to serve as a guide for energy is, indeed, the purpose of most conductors). But the external field has a sufficient radial component to give an inward flow of energy to provide for the loss in the conductor. Only around a conductor of perfect conductivity would the Poynting field be wholly parallel to the conductor.

Quantitatively, the electric field in such a conductor as the one in Fig. 49 is

$$\mathbf{E} = \frac{\iota}{\gamma} \qquad [312]$$

from which, if the conductor radius is r, and the total current I,

$$E_z = \frac{I}{\pi r^2 \gamma} \qquad [313]$$

The magnetic field at the conductor surface is, from equation 207,

$$H_\theta = \frac{I}{2\pi r} \qquad [314]$$

Hence the Poynting field strength just within the conductor surface is

$$P_r = -\frac{I^2}{2\pi^2 \gamma r^3} \qquad [315]$$

where the negative sign indicates that **P** is directed radially inward. The total power entering length l of the conductor is found by multiplying P_r by the surface area $2\pi r l$:

$$\text{Entering power} = \frac{I^2 l}{\pi r^2 \gamma} \qquad [316]$$

But the resistance R of the conductor is (from equation 160):

$$R = \frac{l}{\pi r^2 \gamma} \qquad [317]$$

When this is substituted into equation 316, we find that the energy supplied by the Poynting field is

$$\text{Entering power} = I^2 R \qquad [318]$$

This result is obviously in agreement with the well-known expression for power consumed in resistance. It is here derived by what amounts to an integration of the Poynting field to obtain flow of energy into the conductor. It illustrates a means of computing power. This method is used in a later chapter, where power radiated from a radio antenna is found by integrating the Poynting field over a surface completely surrounding the antenna.

The Poynting field of the plane wave of Fig. 48 is readily determined. From equations 262 and 271:[12]

$$\mathbf{P} = \mathbf{E} \times \mathbf{H} = \mathbf{i} \frac{E_m^2}{\eta} \cos^2(\omega t - \beta x) \qquad [319]$$

This Poynting field is everywhere in the positive x direction, a result that agrees with the obvious direction of flow of energy. It is maximum where $\mathbf{E}$ and $\mathbf{H}$ are greatest, whether they are positive or negative, and it is zero where $\mathbf{E}$ and $\mathbf{H}$ are zero.

Note particularly that the Poynting vector, and therefore the direction of travel of a wave, is boreal to the angle from $\mathbf{E}$ to $\mathbf{H}$. If the fingers of the right hand curve from $\mathbf{E}$ to $\mathbf{H}$ the thumb shows the direction of

---

[12] This is one of the places, previously mentioned, where the exponential form of expression can lead to trouble unless care is used. One can *not* write from equation 276:

$$\mathbf{P} = \mathbf{E} \times \mathbf{H} = (\mathbf{E}_0 \, e^{j\omega t}) \times (\mathbf{H}_0 \, e^{j\omega t}) = (\mathbf{E}_0 \times \mathbf{H}_0) \, e^{j2\omega t}$$

It is, however, quite correct to write

$$\mathbf{P} = \text{Real part of } (\mathbf{E}_0 \, e^{j\omega t}) \times \text{Real part of } (\mathbf{H}_0 \, e^{j\omega t})$$

which is not the same thing. As has been stated, uncertainties can always be cleared away by filling in the phrase "Real part of" (or its customary abbreviation "Re") in the equations where needed. Thus, in general,

$$(\text{Re } e^{jx})(\text{Re } e^{jy}) \neq \text{Re } (e^{jx} e^{jy})$$

But note

$$(\text{Re } e^{jx}) + (\text{Re } e^{jy}) = \text{Re } (e^{jx} + e^{jy})$$

PLANE WAVES

travel of the wave. This very important relation is easily remembered if "$\mathbf{E} \times \mathbf{H}$" is firmly impressed on the mind. Obviously a reversal of the order of these vectors would be ruinous, but the memory can be helped by noting that $\mathbf{E}$ precedes $\mathbf{H}$ as in the alphabet.

PROBLEMS

1. If, in equation 261, $\beta = 1$, what are the wavelength and frequency?

2. The direction of travel of a plane wave is normal to the Z axis and midway between the X axis and the Y axis (at 45 degrees to each of the latter). The wave is polarized with the electric vector parallel to the $X-Y$ plane. Write the necessary defining equations (similar to equations 256 and 257) and introduce them into the wave equation. Find solutions for the components of $\mathbf{E}$ and $\mathbf{H}$, as in equation 259.

3. Prove that $\mathbf{E}$ and $\mathbf{H}$ are perpendicular in direction, identical in form, and proportional in magnitude when the wave is an undefined function $E_y = f(x - vt)$, and not only for the sinusoidal wave of equation 261.

4. Prove that equation 308 is correct. Explain why the relation $\nabla \cdot (\mathbf{B} \times \mathbf{C}) = \mathbf{C} \cdot (\nabla \times \mathbf{B}) - \mathbf{B} \cdot (\nabla \times \mathbf{C})$ is not analogous to $\mathbf{A} \cdot (\mathbf{B} \times \mathbf{C}) = \mathbf{C} \cdot (\mathbf{A} \times \mathbf{B}) = -\mathbf{B} \cdot (\mathbf{A} \times \mathbf{C})$.

5. What is the intrinsic impedance of (a) soft glass, (b) polystyrene, (c) steatite? (Compute from data from a reference book, giving source. Use values applicable at frequency about 10^9; normal temperature.)

6. Consider the equation $V = ZI$ from ordinary alternating-current circuit theory, Z being complex. Show that this is consistent with $v = V_m e^{j\omega t}$ and $i = (I_m e^{-j\phi}) e^{j\omega t}$. If $v = V_m \cos \omega t$ and $i = I_m \cos (\omega t - \phi)$, what is impedance?

7. A wave is defined by $E_x = E_z = 0$, $E_y = 55 e^{j(6283t - \beta z)}$. Write expressions for $\mathbf{E}$, $\mathbf{E}_0$, E_{0y}, and E_m.

8. Compute Γ, α, and β at 1-megacycle frequency for moist earth (see values in Table V, page 184).

9. Compute Γ, α, and β at 1-megacycle frequency for copper, taking conductivity from a reference book (give source) and neglecting dielectric constant.

10. Compute Γ, α, and β at 1-megacycle frequency for hard rubber. Relative dielectric constant is 2.8. Power factor is 0.007 at 1 megacycle.

11. Compare wavelength of a 1-megacycle wave in air, in moist earth (Problem 8), copper (Problem 9), and hard rubber (Problem 10). What is the wavelength in hard rubber if loss is neglected?

12. Show that, as stated, a good approximation for attenuation of a plane wave in a dielectric material is $9.1f \sqrt{\kappa} \times$ (power factor) $\times 10^{-8}$ decibels per meter.

13. Show that a good approximation for attenuation in a semi-conducting medium is $\alpha = \frac{1}{2}\gamma\eta$. How should use of this approximation be restricted to avoid error in α of more than about 5 per cent?

14. If P is power in watts per square meter of surface parallel to the plane of a wave in air, and if E is volts per meter, find n in $P = nE^2$. Check your result with a radio reference book.

15. Show that, in the wave of equations 261 and 269, half the energy is electric and half magnetic. In a medium in which $\kappa = 4$ and $\mu = \mu_0$, what fraction of the total energy is in the electric field?

16. A wire is bent into a loop and the ends are attached to binding posts. A constant voltage is applied between binding posts, and current flows in the wire. Sketch the Poynting vector field about the loop of wire.

17. Considering the possible reception of radio signals by a submerged submarine, compute the distance of travel in sea water that will reduce the field strength of a radio wave to $1/10$ of its initial value. Use $\kappa = 81$ and resistivity $= 1/\gamma = 0.20$ meter-ohm (see Table V, page 184). Compute for signal frequencies of 30 kilocycles per second and 30 megacycles per second.

CHAPTER X

Reflection

Boundary Surfaces. Waves travel in free space or in homogeneous material in simple fashion, but practical problems must always take into account the complicating effects at boundary surfaces. Boundary surfaces lie between regions of different dielectric constant (as between glass and air), or between regions of different conductivity (as between air and copper). There are also boundary surfaces between regions of different permeability which are interesting in general but have little importance as related to waves.

An electric wave is reflected, for example, by a sheet of copper, and reflection is a boundary problem. A wave falling upon a body of di-

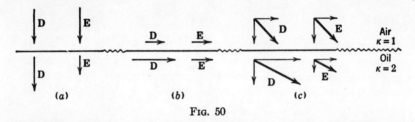

Fig. 50

electric material is partly reflected and partly transmitted. The transmitted portion undergoes refraction; refraction is also a boundary problem. The reception of energy by a receiving antenna is determined by boundary conditions at the surface of the antenna wire. A wave guide makes obvious use of boundary surfaces, and the determination of fields about a transmission line is equally although less evidently a boundary problem. Any conducting wire, as a matter of fact, is important electrically because it offers a boundary surface for an electromagnetic field; but this is an extreme interpretation when applied to, let us say, a lamp cord.

Where an electric field passes from one substance into another of different dielectric constant, as from air into oil, certain conditions must be met. If the field is normal to the surface as in Fig. 50a, there can be no change in the strength of **D** at the surface, for Experiment **IV** has shown that the divergence of **D** is everywhere zero, except where free charge is located, and it is assumed that there is no free charge on the surface under discussion. **D** must therefore be the same just within the oil as it is just above the surface. **E**, correspondingly, is less in the oil; if the relative dielectric constant of the oil is 2 ($\kappa = 2$), **E** is half as great in the oil as in the air.

If an electrostatic field is *tangential* to a dielectric boundary surface, however, as in Fig. 50b, it is **E** that must be continuous in value above and below the surface, for a discontinuity of the tangential field would correspond to curl of the field at the surface, and Experiment II showed that curl of **E** is everywhere zero. Since the tangential component of **E** must be continuous, **D** cannot be continuous but must be inversely proportional to dielectric constant; **D**, in this case, is twice as great in oil as in air.

Experiment II, from which this conclusion was drawn, is limited to the electrostatic state, and in general the curl of **E** is not zero but is $-\partial \mathbf{B}/\partial t$; we must consider whether this changes the conclusion that tangential **E** is continuous through the boundary. First, it must be recognized that an actual discontinuity of **E** at the boundary not only would result in curl, but there would be, at the boundary surface, infinite curl. A discontinuity corresponds to an infinite rate of change of the function. Since there cannot be an infinite value of $\partial \mathbf{B}/\partial t$, it follows that, in the dynamic as well as in the static state, there can be no discontinuity in the tangential **E**.

Finally, if the electric field is at an angle to the boundary surface, it is most readily discussed by considering separately its two components, normal and tangential. Figure 50c shows a field at an angle with each component behaving in accordance with the above rules. As a result, **D** and **E** are parallel in air and in oil, but the angle of both changes as they pass through the surface. The proportion of **D** to **E** in oil is twice that in air, corresponding to $\kappa = 2$.

At a boundary between materials of different permeability, the magnetic field behaves in a similar way: normal **B** is continuous through a boundary to avoid divergence, and the tangential component of **H** is continuous as there cannot be infinite curl.

Conductor as Boundary. At the boundary surface of conducting material, there are two particularly important effects. A conductor is likely to have electric charge on its surface, for charge can flow freely to the surface from any point within the conductor. Also, a conductor is likely to carry current parallel to the surface. Surface charge provides a boundary to terminate electric field, and current provides a boundary for magnetic field.

First let us consider surface charge. Assume a normal component of electric field just outside a conductor; either this field is terminated at the conductor surface by charge which, by equation 122, is $\sigma = D_n$, or else it penetrates into the conductor. However, if field exists in the conductor, there must be a corresponding current in the conductor, $\iota = \gamma E$ (equation 165), and this current will quickly carry electric charge to the surface. Such charge, collecting at the surface, will then terminate the electric field, thereby eliminating penetration of field into the conductor and stopping the flow of current. It follows, as mentioned in Chapter IV, that there can be no static electric field within a conductor. The higher the conductivity, the more quickly the internal field will be eliminated. In a rapidly changing dynamic field, there can be some electric field within a poor conductor but very little in a good conductor. In the limiting case of a perfect conductor (infinite conductivity) there can be no electric field.

The basic principle of shielding a circuit or part of a circuit from electric field by enclosing it within a metal container is evident from the above discussion. The possibility of shielding from magnetic induction will next be considered.

Changing magnetic field induces current in a conductor. First consider a magnetic field outside a conductor and tangential to the surface, as indicated in Fig. 51. The magnetic field is assumed to be increasing with time, so $\partial B/\partial t$ is positive. The material of the block shown in the diagram is a reasonably good conductor of conductivity γ; since it is not a perfect conductor some magnetic field will exist within the material. This magnetic field within the material is increasing with time, and hence there will be an induced electric field (equation 193) and current will flow (equation 165) as shown by the dotted lines. However, on account of the flow of current, the magnetic field will have curl (equation 240) in the direction of the current flow,[1] and this means that B will diminish at increasing depths within the conducting material. Thus, in a conductor that is good but not perfect, an increasing magnetic field will penetrate to some extent; it will induce voltage, and current will flow, and the current will automatically distribute itself in such a way as to weaken the magnetic field and prevent the field from penetrating deeply into the material.

If the tangential magnetic field ceases to increase with time and becomes static, it will no longer induce an electric field in the conductor;

---

[1] By equation 240, $\nabla \times H = \iota + \partial D/dt = \gamma E + \epsilon \partial E/\partial t$. Since γ is high for most metals (of the order of magnitude of 10^7 mhos per meter) and ϵ is low (of the order of 10^{-11}) the term $\partial D/dt$ in equation 240 is negligible in conducting material, even at very high frequency, and approximately $\nabla \times H = \iota$.

CONDUCTOR AS BOUNDARY

current will then quickly cease, and with no current density there will be no curl in the magnetic field. The magnetostatic field will thereafter be the same in the conductor as it would have been in a non-conductor of the same permeability. But it will have required time to establish magnetic field in conducting material. If the external magnetic field is alternating, there is never time for the field to be completely established, and, the higher the frequency, the less the magnetic field that will exist within the conducting material. Also, the higher the frequency, the more the current will be concentrated near the surface of the conductor. This is "skin effect."

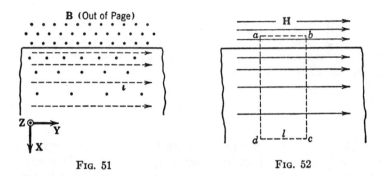

Fig. 51   Fig. 52

The higher the conductivity of the material, in Fig. 51, the greater the current density that results from a given induced electric field. A material of high conductivity will thus have high current density near the surface, but there will be less penetration of magnetic field and current into the material. Skin effect will be more marked. If the extreme case of ideally perfect conducting material ($\gamma = \infty$) is considered, there will be infinite current density just at the surface and no penetration at all.

Current may be thought of as providing a boundary for the magnetic field. The current permits the field to have curl and thus to diminish to zero (or substantially zero), at or near the surface of the conductor.

When current provides a boundary for magnetic field, the total amount of current in the conducting material is not dependent upon conductivity of the material but is determined simply by the strength of the external magnetic field. Distribution of current density depends upon conductivity and frequency, but total current is proportional only to the strength of the external field. To show this, consider Fig. 52 which shows a cross section of the same block of material shown in Fig. 51, in a plane parallel to the field.

The rectangle *abcda* is a path of integration for evaluating the left-hand member of equation 199:

$$\oint \mathbf{H} \cdot d\mathbf{s} = I \qquad [199]$$

(This was originally a magnetostatic equation, but, since we have already considered that displacement current $\partial \mathbf{D}/\partial t$ is negligible within a good conductor, this equation may be considered to be derived from 240.) If the length of the side *ab* is l, integration along *ab* yields, as its contribution to the contour integral, Hl. Integration along *bc* and *da* adds nothing to the integral as these sides are normal to the field. The side *cd* is chosen to be so deep within the material that **H** along *cd* is negligible and integration along *cd* adds nothing to the integral. Hence I, which is total current through the rectangle *abcda*, flowing into the sheet in Fig. 52, equals Hl, or

$$\frac{I}{l} = H \qquad [320]$$

The idea of surface current is helpful in problems concerning good conductors, such as metals, in a high-frequency field. The total current in the metal will be concentrated almost entirely in a region just under the surface, and the amount is easily computed from equation 320.[2] For certain purposes it is satisfactory to assume that the metal is a perfect conductor and that the total current will flow in an infinitely thin sheet right at the surface. This is known to be untrue, and it results in an infinite current density at the surface, but it is a great convenience and a permissible approximation in some problems, such as reflection of waves from a metal surface.

In this discussion we have been speaking of a *tangential* external magnetic field and its relation to what goes on within the metal. The reason for this emphasis on the tangential field is that, in a perfect conductor (or at high frequency), a tangential field is the only kind that can exist to any significant extent. At first it would seem that there might also be a *normal* component of magnetic field, but it must be remembered that, if there were a normal component of **B** just external to the surface, the same normal component would have to appear in the metal just

---

[2] If the surface of the conductor does not happen to be a plane surface, as shown in Fig. 52, equation 320 is nevertheless valid. The path of integration in equation 199 is then so chosen that *ab* is parallel and close to the surface, *bc* and *da* are everywhere normal to **H**, and *cd* is so deep within the material that the magnetic field is negligible.

within the surface because there can be no divergence of **B**. However, there can be no changing field within a perfect conductor because of the process that was described above: the changing magnetic field will produce an electric field, which will produce a current, which will produce an opposing magnetic field, and in material of perfect conductivity the current will flow in whatever amount is necessary exactly to cancel out the penetrating magnetic field. Thus, the magnetic field within the conductor is zero, and the magnetic field normal to its surface must be zero also. From the point of view of the external field, magnetic flux lines approaching a perfect conductor will not penetrate its surface but will bend away and pass by tangentially. With high-frequency fields and ordinary metallic conductors, even though the conductivity is something less than perfect, this statement is a good approximation.

Skin Effect. The limitation of current and magnetic field to a surface layer is commonly called "skin effect." It appears in different guises in such problems as effective resistance of transmission-line conductors, attenuation in wave guides, shielding of one part of a radio circuit from fields produced in another, and penetration of flux into the rotor of an induction motor. These problems are essentially similar and their solutions require appropriate use of the wave equation as applied to propagation within conducting material.

Equation 288 is a solution of the wave equation for electric field in conducting material:

$$E_{0y} = E_m \, e^{\pm \Gamma x} \qquad [288]$$

Since current density in a conductor is proportional to the electric field, a similar equation for current distribution can be written:

$$\iota_{0y} = \iota_m \, e^{\pm \Gamma x} \qquad [288a]$$

In this equation $\Gamma^2 = j\omega\mu(\gamma + j\omega\epsilon)$. However, as mentioned in footnote 1, page 140, when considering fields in ordinary metals, in which conductivity is vastly greater than the dielectric constant, the term $j\omega\epsilon$ within the parentheses is negligible compared with γ unless the frequency is of the order of magnitude of that of visible light. Hence for fields of even the highest radio frequency in metal conductors

$$\Gamma = \sqrt{j\omega\mu\gamma} = \sqrt{\pi f \mu \gamma} \, (1 + j) = \frac{1+j}{\delta}$$

if δ is defined as $1/\sqrt{\pi f \mu \gamma}$.

Equation 288a describes the current distribution shown in Fig. 51 if coordinate axes are so oriented that current ι is in the y direction, **B** is in the z direction, and the X axis points downward. Assume the top surface of the block of conducting material to be at $x = 0$.

It will be seen that the equation permits a choice of sign in the exponent. However, in this block of material, so thick that it is considered to be semi-infinite, the positive sign is discarded, as a term with positive sign would describe a current density increasing without limit deep within the metal. (Actually, such a term is required if the metal is of finite thickness, and there is current penetrating all the way through from the other side.) It is therefore enough to write

$$\iota_{0y} = \iota_m e^{-\Gamma x} = \iota_m e^{-x/\delta} e^{-j(x/\delta)}$$

This equation, which gives the current density at any depth x within the material, relates the current density at any depth to the current density ι_m at the surface, where $x = 0$, but it does not evaluate ι_m. To do this, we must use the information from equation 320 that the total current, obtained by integrating current density from the surface to an infinite depth, equals the magnetic field tangential to the surface. The current carried in a section of the metal block that is one unit wide and that extends from the surface to an unlimited depth (see Fig. 52) is:

$$\frac{I}{l} = \int_0^\infty \iota_{0y}\, dx = \frac{\iota_m}{\Gamma} = H$$

This is an equation with many uses. It leads, for instance, to an expression for power loss when there is skin effect. Basically, the power loss per unit volume is $|\iota^2|/\gamma$. Integration to find the time-average of power loss in a semi-infinite conductor (per unit of surface area) gives $\iota_m^2 \delta/4\gamma$. This is a particularly interesting expression because it is easily shown that, if all the current in the semi-infinite conductor flowed with uniform distribution in a surface layer of thickness δ, the loss would be exactly equal to the actual loss. Hence, by finding the direct-current resistance of a layer of thickness δ at the surface of the conductor, a value called the effective resistance is obtained. It is this that leads to δ being called the "equivalent depth of penetration" of the current, or the "skin depth." *Actually* the current penetrates in appreciable amount to many times the "skin depth," diminishing exponentially.

Although the above derivation of effective resistance is for a semi-infinite block of conducting material, it can be used as a good approximation for any large conductor carrying current at high frequency. Specifically, it is a good approximation if all dimensions of a conductor are many times the "skin depth" δ. As an example, the effective resistance of a round wire at high radio frequency is approximately the direct-current resistance of that part of the wire that is within a distance δ of the surface. Since an exact expression for the effective resistance of a round wire can be computed by finding a solution of the wave equation in a conducting cylinder, it is interesting to compare the approximate and exact results. The approximate result, considering the effective resistance to be the direct-current resistance of a surface layer, is in error by less than 10 per cent if the radius of the wire is as much as six times δ.

Another concept related to skin effect is "surface resistivity" of a conductor. With this meaning, "surface resistivity" is the direct-current resistance (as measured between two opposite edges) of a square of metal 1 meter long and 1 meter wide and δ thick. Hence the "surface resistivity" is $1/\gamma\delta$, and this value will be used in the later chapter on wave guides.

Reflection from a Conductor. When an electric wave is traveling through space, there is an exact balance between the electric and magnetic fields. Half of the energy of the wave, as a matter of fact, is in the electric field and half in the magnetic. If the wave enters some different medium, there must be a new distribution of energy. Whether the new medium is a dielectric material, a magnetic material, a conducting material, or an ionized region containing free charge, there will have to be a readjustment of energy relations as the wave reaches its surface. Since no energy can be added to the wave as it passes through the boundary surface, the only way that a new balance can be achieved is for some of the impinging energy to be rejected. This is what actually happens, and the rejected energy constitutes a reflected wave. Hence one sees reflection of light from a conducting metal surface and from a dielectric glass surface. Often, indeed, the transmitted wave is rapidly absorbed and lost, as when light falls upon porcelain, or wood, or gold; yet, if the porcelain or wood or gold is thin enough, some of the transmitted light will pass through.

The simplest reflection to discuss is that of a plane wave falling upon a perfectly conducting plane surface. This is an extreme case, for in a perfectly conducting material the electric field strength is always zero, and it will be shown that a wave falling upon such a surface is totally reflected.

In Fig. 53 a set of coordinate axes is oriented with the Z axis downward. A perfectly conducting surface coinciding with the X–Y plane will then be horizontal. A polarized plane electromagnetic wave falling on this surface from above is defined by

$$E_{x1} = E_m \cos \beta(vt - z) \qquad [321]$$

This equation, from its similarity to equation 252, is seen to describe a wave traveling in the positive z direction, polarized in the x direction.

146   REFLECTION

But equation 321 gives only the incident wave, and when there is a reflected wave the complete solution of the wave equation must be used. As in equation 259, this includes a wave traveling in the opposite direction, which may have any form and which will be indicated merely as some function $f_2(vt + z)$. The nature of this function must be determined from conditions at the boundary surface.

In space above the X–Y plane, then,

$$E_x = E_m \cos \beta(vt - z) + f_2(vt + z) \qquad [322]$$

The given boundary condition is that the plane surface is perfectly conducting and therefore within the surface $E_x = 0$. Since the tangential

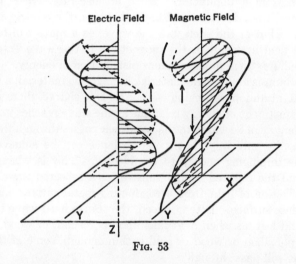

Fig. 53

component of **E** must be continuous through the boundary surface, E_x must be zero just above the surface as well as just below, and at $z = 0$, $E_x = 0$. Substituting this in equation 322 gives

$$0 = E_m \cos \beta(vt) + f_2(vt) \qquad [323]$$

whence

$$f_2(vt) = -E_m \cos \beta(vt) \qquad [324]$$

Changing the argument of the function f_2 from (vt) to $(vt + z)$:

$$f_2(vt + z) = -E_m \cos \beta(vt + z) \qquad [325]$$

It is thus seen that, in order to satisfy the boundary condition at the perfect conductor, the complete electric field of equation 322 must be:

$$E_x = E_m \cos \beta(vt - z) - E_m \cos \beta(vt + z) \qquad [326]$$

Introducing $\beta v = \omega$ (Table III):

$$E_x = E_m \cos (\omega t - \beta z) - E_m \cos (\omega t + \beta z)$$
$$E_y = E_z = 0 \qquad [327]$$

This result could have been obtained as an exponential function, in which form equation 327 is

$$E_x = E_m e^{j(\omega t - \beta z)} - E_m e^{j(\omega t + \beta z)} = E_m (e^{-j\beta z} - e^{j\beta z}) e^{j\omega t}$$
$$E_y = E_z = 0 \qquad [328]$$

The electric component of the field above the reflecting surface is now known; the magnetic component is readily found from Maxwell's equations as was done for a simple wave in equation 269. The result is

$$H_y = \frac{E_m}{\eta} [\cos (\omega t - \beta z) + \cos (\omega t + \beta z)] \qquad [329]$$

or alternatively

$$H_y = \frac{E_m}{\eta} (e^{-j\beta z} + e^{j\beta z}) e^{j\omega t} \qquad [330]$$

and

$$H_x = H_z = 0$$

Both electric and magnetic components are indicated in Fig. 53; each arrow indicates the field strength throughout the corresponding horizontal plane, so electric and magnetic vectors at the same height above the reflecting plane are really coincident. Note particularly that, whereas the electric wave is reflected with reversal of sign so that the electric field at the reflecting surface is always zero, the magnetic field is reflected with unchanged sign and so is doubled at the reflecting plane. The correspondence to traveling waves on transmission lines [3] is more than an analogy, for indeed the wave traveling along a transmission line is a

[3] See *Transient Electric Currents*, H. H. Skilling, McGraw-Hill Book Co., N. Y., 1937.

148   REFLECTION

plane electromagnetic wave, and its reflection is an example of the type of reflection considered here.

Conditions in the conducting plane must be considered, to show that they are consistent with the principles of electrodynamics. The electric field in the reflecting plane must be zero. When the incident wave arrives, current flows in the plane, which can carry unlimited current with zero voltage; the current density is infinite, but its depth of penetration into the plane is zero. Magnetic intensity just above the surface is finite; just below the surface it is zero. This change takes place in zero distance, for it occurs precisely at the surface, and consequently the curl of the magnetic field *at the surface* is infinite. This is not only permissible but necessary if the current density at the surface is infinite. Study of Fig. 53 will show that the directions of current, curl, and magnetic field are in accord with Maxwell's equations.

A heavy line in Fig. 53 shows the sum of the two traveling waves. This resultant is continually changing, but it is not a traveling wave. It is oscillating in magnitude but fixed in space: it is a "standing" wave. The total electric intensity is always zero at the reflecting surface, at a distance of one-half wavelength from the reflecting surface, and at multiples of one-half wavelength. These points are nodes. There are also nodes in the magnetic field, at one-fourth wavelength, three-fourths wavelength, and so on.

It was by detection of these nodes in front of a reflecting sheet of zinc that Hertz first proved the existence of electromagnetic waves. He explored the field with a wire loop about a foot in diameter, with the ends of the loop separated by a very minute distance; electric sparks across this small gap indicated an induced electromotive force in the loop, and the absence of sparks indicated that his loop of wire was located at a node. Since there can be nodes only if there are waves, Maxwell's theory was proved true.

The standing waves are best seen in mathematical form by changing equation 328 to

$$E_x = -2E_m \frac{e^{j\beta z} - e^{-j\beta z}}{2j} j\, e^{j\omega t} \qquad [331]$$

$$= 2E_m \sin \beta z (-j\, e^{j\omega t}) \qquad [332]$$

This expression really means

$$E_x = \text{Real part of } [2E_m \sin \beta z (-j\, e^{j\omega t})] \qquad [333]$$

To find the real part, the exponential is expanded as on page 125, using

$-j\,e^{j\omega t} = -j(\cos \omega t + j \sin \omega t) = \sin \omega t - j \cos \omega t$. Retaining only the real part [4] leaves

$$E_x = 2E_m \sin \beta z \sin \omega t \qquad [334]$$

(This result could have been obtained from equation 327 by trigonometric substitution, but less easily.)

Equation 334 gives the electric field in space above the reflecting plane. The magnetic field is found from equation 330 to be

$$H_y = \frac{2E_m}{\eta} \cos \beta z \cos \omega t \qquad [335]$$

These last two equations describe waves that do not travel but stand in a fixed position along the Z axis and pulsate. Maximum pulsation occurs at "loops," and zero pulsation at "nodes." It will be seen that loops of electric field occur where $\sin \beta z = \pm 1$; nodes, where $\sin \beta z = 0$. Where there are loops of electric field, there are nodes of magnetic field. Also, the standing waves of magnetic and electric field pulsate out of phase in time, so that when the magnetic field is everywhere zero, the electric field is everywhere maximum, and *vice versa*. Thus the standing wave has a very different appearance from a traveling wave, although it is actually nothing more than the sum of two traveling waves.

This discussion has considered reflection at a plane of infinite conductivity. At a surface of finite conductivity, such as copper, conditions will be nearly the same. A small amount of energy but not much will be carried by a weak transmitted wave a short distance into the copper, while most of the incident energy will be reflected as from a perfect conductor.

Dielectric Reflection. When a wave falls onto the surface of a block of dielectric material, there is partial reflection and partial transmission of the incident energy. Whether energy passing into the dielectric material is transmitted through that material without rapid attenuation depends on the characteristics of the material; attenuation results if the material fails to be perfectly non-conducting, or if there is loss associated with each reversal of the electric field ("dielectric hysteresis"). In microwave transmission, the latter is much the more important factor,

[4] In changing expressions from the trigonometric to the exponential form, there are two especially useful relations that have now been developed. They are

$$\cos x = \text{Real part of } e^{jx}$$

$$\sin x = \text{Real part of } -j\,e^{jx}$$

These will be used frequently.

150   REFLECTION

for dielectric loss increases approximately with the frequency. At light-wave frequencies, energy absorption within the atoms is important, and thus many dielectrics that transmit radio waves are opaque to light.

As an example, let us consider a wave in free space, or air, falling normally upon a large block of dielectric material with constant ϵ. See Fig. 54. In the following work a subscript 1 will indicate the incident wave, 2 the reflected wave, and 3 the transmitted wave. The same orientation with respect to axes will be used as in Fig. 53. It is hardly necessary to prove again that the reflected wave will have the same

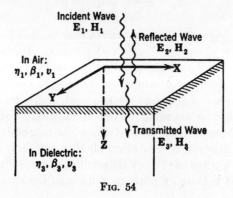

Fig. 54

functional form as the incident wave, and this will be accepted. Then in air

$$E_{x(\text{air})} = E_{m1}\, e^{j(\omega t - \beta_1 z)} + E_{m2}\, e^{j(\omega t + \beta_1 z)} \qquad [336]$$

and in the dielectric material

$$E_{x(\text{diel})} = E_{m3}\, e^{j(\omega t - \beta_3 z)} \qquad [337]$$

Since the tangential component of electric field is continuous at the boundary surface, we know that, at $z = 0$,

$$E_{x(\text{air})} = E_{x(\text{diel})} \qquad [338]$$

and equating 336 and 337 at the surface, where $z = 0$, gives

$$E_{m1} + E_{m2} = E_{m3} \qquad [339]$$

As in previous examples, the magnetic field of each traveling wave is normal to the electric field and normal to the direction of travel of the wave.

$$H_{y(\text{air})} = H_{m1}\, e^{j(\omega t - \beta_1 z)} + H_{m2}\, e^{j(\omega t + \beta_1 z)} \qquad [340]$$

$$H_{y(\text{diel})} = H_{m3}\, e^{j(\omega t - \beta_3 z)} \qquad [341]$$

It follows from Maxwell's equations, as in earlier examples, that E and H are related in magnitude by the intrinsic impedance of the medium through which the wave is traveling.

$$E_{m1} = \eta_1 H_{m1} \qquad E_{m2} = -\eta_1 H_{m2} \qquad E_{m3} = \eta_3 H_{m3} \qquad [342]$$

The negative sign for the reflected component of the wave indicates that (considering the direction of travel, and the boreal relation of E, H, and velocity), when the electric component is positive (along the X axis), the magnetic component (along the Y axis) is negative. At the boundary surface the tangential component of **H** is continuous, so, when $z = 0$, $H_{y(\text{air})} = H_{y(\text{diel})}$ and

$$H_{m1} + H_{m2} = H_{m3} \qquad [343]$$

or

$$\frac{E_{m1}}{\eta_1} - \frac{E_{m2}}{\eta_1} = \frac{E_{m3}}{\eta_3} \qquad [344]$$

From equations 339 and 344 the relative magnitudes of incident, reflected, and transmitted waves are found:

$$E_{m2} = \frac{\eta_3 - \eta_1}{\eta_3 + \eta_1} E_{m1} \qquad E_{m3} = \frac{2\eta_3}{\eta_3 + \eta_1} E_{m1} \qquad [345]$$

$$H_{m2} = -\frac{\eta_3 - \eta_1}{\eta_3 + \eta_1} H_{m1} \qquad H_{m3} = \frac{2\eta_1}{\eta_3 + \eta_1} H_{m1} \qquad [346]$$

The fraction $(\eta_3 - \eta_1)/(\eta_3 + \eta_1)$ is called the **reflection coefficient**.

It is more interesting to consider a special case than to attempt any discussion of these general results. Let us suppose the dielectric material of this example to have a relative dielectric constant of 4 (Bakelite, perhaps) so that $\epsilon_3/\epsilon_1 = 4$. Since $\eta = \sqrt{\mu/\epsilon}$, and μ is practically the same for air and any dielectric material, we find $\eta_1/\eta_3 = 2$. For this special case, equations 345 and 346 become

$$E_{m2} = -\tfrac{1}{3} E_{m1} \qquad E_{m3} = \tfrac{2}{3} E_{m1} \qquad [347]$$

$$H_{m2} = \tfrac{1}{3} H_{m1} \qquad H_{m3} = \tfrac{4}{3} H_{m1} \qquad [348]$$

Note that E and H in the reflected wave have the same ratio of magnitude as in the incident wave, and the reflected wave has one-third the amplitude of the incident wave. The reflection coefficient is $-\tfrac{1}{3}$.

The transmitted wave, being in a different medium, has a different ratio of E to H as determined by the value of η.

In this simple wave it is safe to consider power of each component wave individually. The Poynting vector of power per unit area at $z = 0$ in the incident wave is

$$\mathbf{P}_1 = \mathbf{E}_1 \times \mathbf{H}_1 \qquad [349]$$

This is entirely in the z direction, and its scalar magnitude is

$$\begin{aligned} P_1 &= E_{x1} H_{y1} \\ &= E_{m1} \cos \omega t \; H_{m1} \cos \omega t \\ &= E_{m1} H_{m1} \cos^2 \omega t \end{aligned} \qquad [350]$$

Energy in the incident wave is thus arriving at the dielectric surface as a double-frequency pulsation that is always positive. The maximum value, P_{m1}, is

$$P_{m1} = E_{m1} H_{m1} = \frac{E_{m1}^2}{\eta_1} \qquad [351]$$

The maximum value of power in the reflected wave is similarly

$$\begin{aligned} P_{m2} &= E_{m2} H_{m2} \\ &= \left(\frac{1}{3} E_{m1}\right)\left(\frac{1}{3} H_{m1}\right) = \frac{1}{9} \frac{E_{m1}^2}{\eta_1} \end{aligned} \qquad [352]$$

and in the transmitted wave

$$\begin{aligned} P_{m3} &= E_{m3} H_{m3} \\ &= \left(\frac{2}{3} E_{m1}\right)\left(\frac{4}{3} H_{m1}\right) = \frac{8}{9} \frac{E_{m1}^2}{\eta_1} \end{aligned} \qquad [353]$$

It is seen that $\frac{1}{9}$ of the incident energy is rejected at the surface and is carried away by the reflected wave, while $\frac{8}{9}$ of the energy passes on into the dielectric material.

We may now inquire whether the electric field in the region above the dielectric reflecting surface can be expressed as a standing wave, as it was with a perfectly conducting reflecting surface. The answer is that it cannot, but it can be expressed as the sum of a traveling wave and a standing wave. If the reflection coefficient is near unity (this corresponds to an extreme change in ϵ, μ, or γ at the reflecting surface), the reflected wave is nearly as large as the incident wave, and the total field has a standing-wave appearance. See Fig. 55.

REFLECTION FROM A SEMI-CONDUCTOR

If the reflection coefficient is near unity, the maximum (or rms effective) field strength will vary as a function of distance above the surface, with marked nodes and loops. The **standing-wave ratio**, which is defined as the ratio of the field strength at a loop to the field strength at a node, will be large.

If the reflection coefficient is near zero (this corresponds to a small change of constants at the reflecting surface), there is very little reflected wave, and the total field above the reflecting surface is only slightly modified from a pure traveling wave. The maximum (or rms effective)

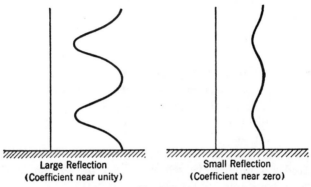

Large Reflection (Coefficient near unity)   Small Reflection (Coefficient near zero)

Fig. 55

field strength will be almost independent of position above the reflecting surface. The **standing-wave ratio** in this case is near unity.

Reflection from a Poor Conductor. If the resistivity of a conductor is so great that there can be a considerable electric field within the material, or (what is really the same thing) if the conductivity of a dielectric is enough to permit appreciable current, reflection from the surface shows some interesting results. Propagation in such a medium, which is neither a perfect dielectric nor a perfect conductor, was considered in Chapter IX. The complex propagation factor Γ was introduced, and the intrinsic impedance η was found to be complex when there is conductivity.

Our discussion of dielectric reflection applies without change to reflection from the surface of a poor conductor if we use the appropriate value of Γ for the propagation factor in the medium, and the proper complex value of η. Thus, if a wave in air falls on a poor conducting material, as in Fig. 54, we write Γ_3 instead of $j\beta_3$ in describing the transmitted wave. Equation 337 becomes

$$E_{x(\text{diel})} = E_{m3}\, e^{j\omega t - \Gamma_3 z} \qquad [354]$$

and equation 341 must be written

$$H_{y(\text{diel})} = H_{m3}\, e^{j\omega t - \Gamma_3 z} \quad [355]$$

Equation 342 remains just as before,

$$E_{m1} = \eta_1 H_{m1} \qquad E_{m2} = -\eta_1 H_{m2} \qquad E_{m3} = \eta_3 H_{m3} \quad [342]$$

It must be recognized that η_3 is now a complex number, and from equation 301:

$$\eta_3 = \sqrt{\dfrac{\mu_3}{\epsilon_3\left(1 + \dfrac{\gamma_3}{j\omega\epsilon_3}\right)}} \quad [356]$$

Because the electric and magnetic fields of the transmitted wave are not in time phase with each other (η_3 being complex and representing a phase shift), it follows that the reflected wave cannot be exactly in phase with the incident wave at the boundary surface and still satisfy equations 339 and 343.

As an example of the phase shift that occurs when a wave is reflected, let us consider a radio wave reflected from the surface of the earth. The following constants are typical of moist earth.

$$\gamma = 10^{-2} \text{ mho per meter}$$

$$\kappa = 25$$

$$\epsilon = \kappa\epsilon_0 = 220 \times 10^{-12}$$

$$\mu = \mu_0 = 1.26 \times 10^{-6}$$

The wave, with a frequency of 10 megacycles per second, and hence an ω of 6.28×10^7, is assumed to fall normally upon the earth's surface. The intrinsic impedance of the earth is

$$\eta_3 = \sqrt{\dfrac{1.26 \times 10^{-6}}{220 \times 10^{-12}(1.24\underline{/-35.9°})}} = 68.0\underline{/18.0°} \text{ ohms}$$

It is interesting to compare this with the intrinsic impedance of free space, 377 ohms. The difference is largely due to the high dielectric constant of the moist earth, rather than the conductivity; the denominator under the radical is changed by a factor of 25 because of dielectric constant, and only by 1.24 because of conductivity. However, the conductivity gives the 18-degree angle to the intrinsic impedance, indicating that H lags 18 degrees (of time) behind E in the earth.

OBLIQUE REFLECTION

From the practical point of view we wish to know the reflection factor, for this will tell the magnitude and phase of the reflected wave, which is ordinarily the useful component. From equation 345,

$$\frac{E_{m2}}{E_{m1}} = \frac{68.0/18.0° - 377}{68.0/18.0° + 377} = -0.705/-6.6°$$

This says the reflected wave is about 0.7 of the incident wave in amplitude (and therefore about half in energy). The reflected wave is almost opposite in phase to the incident wave at the reflecting surface

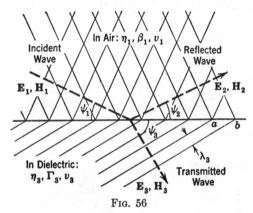

Fig. 56

but, because of the earth's conductivity, it lags 6.6 degrees behind ideal phase reversal.

The example shows that, at 10 megacycles, even moist earth acts more like a poor dielectric than a conductor. At very low radio frequencies, on the other hand, the earth acts more like a poor conductor than a dielectric.

Oblique Reflection. The majority of practical problems are concerned not with normal reflection but with reflection at an angle. The general problem of a wave of any polarization, traveling in any material and falling at any angle onto the surface of any other material, can become fairly complicated. In principle, however, oblique reflection does not differ much from normal reflection.

Let us consider one of the more important examples. A wave in air (or other non-conducting medium) falls at an angle onto the surface of a semi-conducting material such as the earth. Let us assume horizontal polarization (the electric vector parallel to the reflecting surface). This is assumed because of the practical interest in earth-reflected waves in the very high frequency range, for which horizontal polarization is usual.

In Figure 56 the heavy dash line shows the direction of travel of each wave, and the light lines show the advancing wave-crests of the plane waves. In the statement of the problem, only the incident wave is known; the reflected and transmitted waves are to be found.

The solution is based on the familiar boundary condition: at the reflecting surface the tangential components of both E and H must be continuous. This is expressed in the equations

$$E_1 + E_2 = E_3 \qquad [357]$$

$$H_{t1} + H_{t2} = H_{t3} \qquad [358]$$

The subscript t indicates the tangential component at the surface,

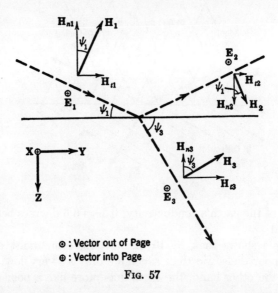

⊙ : Vector out of Page
⊕ : Vector into Page

Fig. 57

as in Fig. 57. Since the incident wave is horizontally polarized, the E component is entirely tangential to the surface, and no subscript t is needed (although in a more general case the equation would be $E_{t1} + E_{t2} = E_{t3}$).

It is hardly necessary to go through detailed analysis to see that, for the sum of the incident and reflected waves to be equal to the transmitted wave all over the boundary surface, the waves must all be of the same frequency. Also, they must all be moving along the surface in the same direction, from left to right in Fig. 56. Also, the distance along the surface from wave-crest to wave-crest must be the same for all three waves. (This is the distance a–b in the figure, for each component wave.) If any of these conditions were not met, equations 357 and 358 could not be satisfied.

Because the transmitted wave travels more slowly than the incident and reflected waves (assuming a higher dielectric constant in the earth), and the wavelength is therefore less, the only way wave-crests can be matched at the surface is to adjust the angle of the transmitted wave. It is evident that, since λ, the wavelength, is the distance between wave-crests as shown in Fig. 56,

$$\text{Distance } a\text{-}b = \frac{\lambda_3}{\cos \psi_3} = \frac{\lambda_1}{\cos \psi_1} = \frac{\lambda_2}{\cos \psi_2} \qquad [359]$$

The angle ψ_1 is the angle of the incident wave shown in the figure (this is the angle usually considered in connection with reflection of radio waves, although its complement is the angle defined in optics as the angle of incidence). Similarly, ψ_2 and ψ_3 are the complements of the angles of reflection and refraction.

Since the incident and reflected waves travel in the same medium, $\lambda_1 = \lambda_2$ and $\psi_1 = \psi_2$. That is, the angle of incidence must equal the angle of reflection. The angle of refraction, however, is computed from equation 359:[5]

$$\frac{\cos \psi_1}{\cos \psi_3} = \frac{\lambda_1}{\lambda_3} = \frac{v_1}{v_3} = \frac{\beta_3}{\beta_1} \qquad [360]$$

Relations in the wave just at the reflecting surface are shown in Fig. 57. As in all plane waves, E and H are related by the intrinsic impedance:

$$E_1 = \eta_1 H_1 \qquad E_2 = \eta_1 H_2 \qquad E_3 = \eta_3 H_3 \qquad [361]$$

From these, and Fig. 57:

$$H_{t1} = \frac{E_1}{\eta_1} \sin \psi_1 \qquad H_{t2} = -\frac{E_2}{\eta_1} \sin \psi_1 \qquad H_{t3} = \frac{E_3}{\eta_3} \sin \psi_3 \qquad [362]$$

The negative sign indicates that, when the tangential component of H_2 is positive, E_2 is negative; this follows from the choice of coordinates and the direction of propagation, as in the figure. Substituting into equation 358:

$$\frac{E_1}{\eta_1} \sin \psi_1 - \frac{E_2}{\eta_1} \sin \psi_1 = \frac{E_3}{\eta_3} \sin \psi_3 \qquad [363]$$

[5] The ratios of equation 360 are the "index of refraction" of optics if the incident wave is in vacuum. For a non-conducting material this is $\sqrt{\mu_3 \epsilon_3 / \mu_0 \epsilon_0}$, and for a non-magnetic non-conducting material (the usual case) it is $\sqrt{\epsilon_3 / \epsilon_0}$ or $\sqrt{\kappa_3}$. Optical values are not comparable with radio values, however, as ϵ is a function of frequency at optical frequencies. Nevertheless, the sparkle of a diamond may be attributed to its high dielectric constant.

Simultaneous solution of this and equation 357 gives the reflection factor:

$$\frac{E_2}{E_1} = \frac{\eta_3 \sin \psi_1 - \eta_1 \sin \psi_3}{\eta_3 \sin \psi_1 + \eta_1 \sin \psi_3} \qquad [364]$$

Naturally, the reflection factor for normal reflection is a special case of equation 364, resulting when $\psi = 90$ degrees. For a glancing wave, ψ_1 approaches zero but ψ_3 does not, and E_2 therefore approaches $-E_1$.

The reflection factor of equation 364 (and, indeed, the entire discussion) are valid whether the refracted wave enters material that is conducting or non-conducting. It applies, therefore, to an actual earth as well as to an ideal earth. The only difference is that η_3 for an actual earth is complex, as determined from equation 356, and β_3 in equation 360 is the imaginary component of Γ from equation 294.

It is interesting and of practical importance (as will be seen in the later discussion of antennas) to note that, when a horizontally polarized wave is reflected from the earth, the *phase* of the reflected wave will not differ more than a few degrees from complete reversal of the incident wave. This is shown by the example of the preceding section, together with the deduction from equation 364 that normal incidence gives maximum phase difference as a result of earth conductivity, and that reflection at any lesser angle will come even closer to giving complete reversal of phase. Also, the *magnitude* of the reflected wave will not be greatly different from the magnitude of the incident wave. In the example computed it was seven-tenths; it may be somewhat more or less than this for other examples of normal incidence; it approaches equality with the incident wave for glancing incidence. These considerations help justify the assumption that is often made that at the surface of the earth a horizontally polarized wave is totally reflected with perfect reversal of phase.

A vertically polarized wave can be handled by a mathematical treatment similar to that which we have outlined for the horizontally polarized wave. A wave polarized at any angle can be analyzed into horizontally and vertically polarized components. Thus, with patience, the approach that has been suggested here can be applied to most reflection problems.

PROBLEMS

1. An electric field passes from air into oil ($\kappa = 2$). The field in air is at an angle of 45 degrees to the surface of the oil. What is the angle between the surface and the field in the oil? Find this angle for both **D** and **E**.

2. Is electric field always normal to the surface at a metal surface? Explain the relations of internal and external field in Fig. 44, page 101.

3. Prove that B_n, the component of **B** normal to a perfectly conducting surface, is either zero or static.

4. Prove that $\partial B_{\tan}/\partial n = 0$. $B_{\tan}$ is the component of **B** tangential to a perfectly conducting surface in non-conducting space, and the derivative is taken normal to that surface. (Prove for a plane surface. Can the proof be extended to apply at any surface?)

5. Magnetic field H is tangential to the surface of a semi-infinite conductor. Compute current density at distances x within the material such that $x/\delta = 1, 2,$ and 3. Show the relation of phase and magnitude of these current densities by plotting them as complex vectors radiating from a common origin. On the same diagram show the phase of the total current I, and the surface field **H**.

6. Find the magnetic field at any depth x within a semi-infinite conductor, knowing the magnetic field tangential to the surface. Plot vectors of H at various depths, as was done for current density in Problem 5; use $x/\delta = 0, 1, 2,$ and 3. What can be said about H at a distance δ *above* the conductor surface? Compare in phase with the current-density vectors of Problem 5.

7. Show that the time-average of power loss in a sheet of conducting material 1 meter square and of thickness δ, carrying current parallel to one edge of the square, the current being uniformly distributed and the total current being $(\iota_m/|\Gamma|) \cos \omega t$, is $\iota_m^2 \delta/4\gamma$.

8. What value of t in equation 327 would give agreement with Fig. 53?

9. If the effective (rms) value of the incident electric field in the wave of Fig. 53 is 10 millivolts per meter, what current flows in the conducting $X-Y$ plane? Find the effective (rms) value. Assume the wave is in air.

10. A wave in air falls normally on a paraffin surface ($\kappa = 2.2$). Find the reflection coefficient and the standing wave ratio.

11. Repeat Problem 10 for a wave traveling in paraffin, falling on a paraffin-air surface.

12. Repeat Problem 10 for a wave in air falling normally on a water surface ($\kappa = 80$), the frequency being so high that the water may be considered a perfect dielectric.

13. Sketch curves similar to Fig. 55 showing the results of Problems 10, 11, and 12.

14. Following equation 356, an example is given of reflection from moist earth. Write the complete equations for electric and magnetic fields in space above the surface, and below. Give numerical values where known, and include time relations.

15. Find the reflection factor from moist earth (as in the example following equation 356) for a wave with a frequency of 10,000 megacycles per second. Repeat for a frequency of 10,000 cycles per second.

16. Find the reflection factor, as in the example following equation 356, for the same wave from drier earth, with $\kappa = 10$, $\gamma = 0.004$ mho per meter.

17. Draw a diagram similar to Fig. 56 showing wave fronts as a wave in a dielectric material falls obliquely on the boundary plane between dielectric and air. Consider the critical angle of incidence for which wave fronts in air are normal to the boundary surface, and the relation to total reflection.

18. Derive an equation, similar in form to 364, for the reflection factor for a vertically polarized wave (more precisely, a wave with the magnetic vector parallel to the reflecting surface). Show that, if reflection is from the surface of a perfect dielectric, there is an angle of incidence for which the reflection factor is zero (the Brewster angle).

CHAPTER XI

Radiation

Electrodynamic Potentials. It is convenient to solve electrostatic problems in terms of the electrostatic scalar potential field, for where there is no electric charge and the dielectric material is homogeneous it is necessary only to solve Laplace's equation

$$\nabla^2 V = 0 \qquad [119]$$

In magnetostatic problems there is similar convenience in the use of the magnetic vector potential, for in homogeneous material that carries no current we merely solve

$$\nabla^2 \mathbf{A} = 0 \qquad [365]$$

For dynamic problems, however, including work with waves, these static equations are incomplete. It will be shown that the full dynamic equations to apply in free space or in homogeneous material in which there is no charge or current are

$$\nabla^2 V - \mu\epsilon \frac{\partial^2 V}{\partial t^2} = 0 \qquad [366]$$

$$\nabla^2 \mathbf{A} - \mu\epsilon \frac{\partial^2 \mathbf{A}}{\partial t^2} = 0 \qquad [367]$$

These equations obviously reduce to the two static equations if the potentials are unchanging with time.

Comparison with equation 249 shows that 366 and 367 are wave equations; that is, both potentials are propagated through space with the velocity $v = 1/\sqrt{\mu\epsilon}$. Therefore, when a moving charge or a changing current initiates a change in potential, that change will not affect conditions at a distant point until a lapse of time has permitted a traveling wave to reach the distant point.

In deriving the dynamic potentials, we begin with Maxwell's equations instead of the simpler relations of electrostatics and magnetostatics. The derivation then proceeds the same, step by step. In

ELECTRODYNAMIC POTENTIALS

TABLE IV

Static Potentials		Dynamic Potentials	
$\nabla \times \mathbf{E} = 0$	[111]	$\nabla \times \mathbf{E} = -\dfrac{\partial \mathbf{B}}{\partial t}$	[193]
$\nabla \times \mathbf{H} = \iota$	[203]	$\nabla \times \mathbf{H} = \iota + \dfrac{\partial \mathbf{D}}{\partial t}$	[240]
$\nabla \cdot \mathbf{E} = \dfrac{\rho}{\epsilon}$	[116]	$\nabla \cdot \mathbf{E} = \dfrac{\rho}{\epsilon}$	[116]
$\nabla \cdot \mathbf{B} = 0$	[196]	$\nabla \cdot \mathbf{B} = 0$	[196]

Since

$$\nabla \cdot \mathbf{H} = \frac{1}{\mu} \nabla \cdot \mathbf{B} = 0$$

we can let

$$\mathbf{H} = \nabla \times \mathbf{A} \qquad [211]$$

Since from 111

$$\nabla \times \mathbf{E} = 0 \qquad [111]$$

we can let

$$\mathbf{E} = -\nabla V \qquad [112]$$

Then from 116

$$\nabla \cdot \nabla V = \nabla^2 V = -\frac{\rho}{\epsilon} \qquad [118]$$

From 203

$$\nabla \times \mathbf{H} = \iota$$

$$\nabla \times \nabla \times \mathbf{A}$$

$$= \nabla(\nabla \cdot \mathbf{A}) - \nabla^2 \mathbf{A} = \iota \qquad [213]$$

But we now stipulate

$$\nabla \cdot \mathbf{A} = 0 \qquad [214]$$

Hence

$$\nabla^2 \mathbf{A} = -\iota \qquad [215]$$

and as above

$$\nabla^2 V = -\frac{\rho}{\epsilon} \qquad [118]$$

Since

$$\nabla \cdot \mathbf{H} = \frac{1}{\mu} \nabla \cdot \mathbf{B} = 0$$

we can let

$$\mathbf{H} = \nabla \times \mathbf{A} \qquad [211]$$

Then from 193

$$\nabla \times \mathbf{E} = -\mu \frac{\partial}{\partial t} \nabla \times \mathbf{A} \qquad [368]$$

Since from 368

$$\nabla \times \left(\mathbf{E} + \mu \frac{\partial \mathbf{A}}{\partial t} \right) = 0 \qquad [369]$$

we can let

$$\mathbf{E} + \mu \frac{\partial \mathbf{A}}{\partial t} = -\nabla V \qquad [370]$$

whence

$$\mathbf{E} = -\left(\nabla V + \mu \frac{\partial \mathbf{A}}{\partial t} \right) \qquad [371]$$

Then from 116

$$\nabla \cdot \left(\nabla V + \mu \frac{\partial \mathbf{A}}{\partial t} \right)$$

$$= \nabla^2 V + \mu \frac{\partial}{\partial t} \nabla \cdot \mathbf{A} = -\frac{\rho}{\epsilon} \qquad [372]$$

From 240

$$\nabla \times \mathbf{H} - \frac{\partial \mathbf{D}}{\partial t} = \iota$$

$$\nabla \times \nabla \times \mathbf{A} - \epsilon \frac{\partial \mathbf{E}}{\partial t} \qquad [373]$$

$$= \nabla(\nabla \cdot \mathbf{A}) - \nabla^2 \mathbf{A} + \epsilon \frac{\partial}{\partial t} \left(\nabla V + \mu \frac{\partial \mathbf{A}}{\partial t} \right)$$

$$= \nabla \left(\nabla \cdot \mathbf{A} + \epsilon \frac{\partial V}{\partial t} \right) - \nabla^2 \mathbf{A} + \mu\epsilon \frac{\partial^2 \mathbf{A}}{\partial t^2}$$

$$= \iota$$

But we now stipulate

$$\nabla \cdot \mathbf{A} = -\epsilon \frac{\partial V}{\partial t} \qquad [374]$$

Hence

$$\nabla^2 \mathbf{A} - \mu\epsilon \frac{\partial^2 \mathbf{A}}{\partial t^2} = -\iota \qquad [375]$$

and from 374 and 372

$$\nabla^2 V - \mu\epsilon \frac{\partial^2 V}{\partial t^2} = -\frac{\rho}{\epsilon} \qquad [376]$$

Table IV, the familiar derivation of the static potential equations is given on the left, and the parallel derivation of dynamic potentials on the right. It is assumed that μ and ϵ are constant.

Except for the changes introduced by using Maxwell's equations, the dynamic equations are similar to the static equations until a stipulation is made regarding the divergence of the vector potential. It will be remembered from the discussion in Chapter VI that any divergence can be specified for **A**, and the most convenient one is chosen; for static fields we let $\nabla \cdot \mathbf{A} = 0$. For dynamic fields the relation of equation 374 is specified. When this is done, equations 375 and 376 result.

If a solution for these differential equations, 375 and 376, could easily be found, it would be very useful, for, if current and charge were known, **A** and V could then be found, and from them **E** and **H** by equations 371 and 211. The possibility of finding solutions for equations 375 and 376 can be considered as a single problem, for these equations are of the same form. Some of the characteristics of the solution are evident. An example will help to show what is meant.

Let us consider a simple electromagnetic disturbance. A pulse of current flows in a short piece of wire (as in an antenna); charge flows suddenly from one end of the wire to the other. Before and after the pulse, conditions are static, and the solutions of equations 375 and 376 must be the static-potential solutions of 215 and 118, which are [1]

$$\mathbf{A} = \frac{1}{4\pi} \int \frac{\iota}{r} d\mathcal{V} \qquad V = \frac{1}{4\pi\epsilon} \int \frac{\rho}{r} d\mathcal{V} \qquad [146]$$

These static solutions must apply both before the disturbance starts and after it dies away, but during the disturbance, as seen from equations 366 and 367, changes of **A** and V are produced near the antenna and are propagated through space as traveling waves. (It will be noted that, in this example, equations 375 and 376 reduce to wave equations everywhere except within the antenna.)

We therefore know that the initial static-potential field is changed to the final static-potential field by a disturbance in the form of a traveling wave. The potential at every point in space is affected by the disturbance, but the more distant points will not be affected until after a lapse of time proportional to distance. The disturbance travels with a speed v (previously defined as $v = 1/\sqrt{\mu\epsilon}$) and, hence, if the distance from

---

[1] In this chapter "$\mathcal{V}$" is used as the symbol for volume, to avoid confusion with the use of v for velocity.

the antenna to a point in space is r, the potential at that point will change after a time delay of r/v.

To express the same idea a little differently, at a given time t the potentials **A** and V at a point some distance r from the source of the disturbance will be determined, not by the current and charge in the antenna at that particular time t, but by the current and charge that existed at a slightly earlier time $(t - r/v)$. (This is similar to saying that when one looks at a star several light-years distant he does not see the star as it is today but as it was some years ago.)

Using functional notation, if charge density is a function of time it is written $\rho(t)$; this indicates any function of the variable t. If ρ changes with time in the same manner as $\rho(t)$ but a little later, the variable is written $(t - t_0)$ instead of t; this introduces a time delay of t_0. The delayed charge-density function is thus written $\rho(t - t_0)$.

With this notation, equation 146 for the electrostatic potential is written:

$$V = \frac{1}{4\pi\epsilon} \int \frac{\rho(t)}{r} d\mathbb{v} \qquad [377]$$

where the t in parentheses merely indicates that ρ is a function of time. Now, in view of the time delay to be expected in the corresponding expression for the dynamic potential, our discussion indicates that the solution of equation 376 might reasonably be

$$V = \frac{1}{4\pi\epsilon} \int \frac{\rho\left(t - \dfrac{r}{v}\right)}{r} d\mathbb{v} \qquad [378]$$

thereby making the potential V at time t dependent on charge distribution at the earlier time $(t - r/v)$. A solution for equation 375 would correspondingly be

$$\mathbf{A} = \frac{1}{4\pi} \int \frac{\iota\left(t - \dfrac{r}{v}\right)}{r} d\mathbb{v} \qquad [379]$$

These seem reasonable, but they have been partly based on speculation. As with most differential equations, the proof is rather easily obtained, once a solution is guessed, by substituting the proposed solution back into the differential equation. Thus 378 is substituted into equation

376 which is thereby [2] reduced to an identity. The similarity of equation 375 then shows that its solution is truly 379.

These dynamic potentials are often called **retarded potentials** because of the time delay involved.

It is not always necessary to use retarded potentials even in dynamic problems. When times are long and distances are short, the retarded potentials are indistinguishable from the static potentials. Mathematically, if r is small compared to vt, a function of $(t - r/v)$ is negligibly different from a function of t. Then the simpler static equations can be used even though fields are slowly changing, and this condition is called the **quasi-stationary state**. All problems at power frequencies are quasi-stationary except those that deal with long transmission lines. The time delay in the propagation of a magnetic field within a generator, for instance, is negligible.

Radiation. The electrical phenomenon that is most essentially a dynamic problem is radiation of waves from an antenna. When a short radio antenna was considered from the quasi-stationary point of view in Chapter VI, there was no suggestion of radiation of energy, for radiation is the factor that the quasi-stationary solution overlooks. So, to determine radiation from an antenna, we will seek a solution for the dynamic vector potential as given by equation 379. That is the purpose for which the concepts of vector potential and retarded potential have been introduced.

[2] The substitution of V from equation 378 into equation 376 is not entirely straightforward, however. Direct substitution of equation 378 into equation 376 leads to difficulty in determining the Laplacian of ρ/r at points at which r may be zero. Since r is the distance from an element of charge to the point at which potential is being determined, it can be zero only when potential is being determined at a point at which charge is located. To avoid the difficulty, space is divided into two regions: one so close to the point at which potential is being determined that quasi-stationary conditions apply and $\nabla^2 V = -\rho/\epsilon$; the other containing all the rest of space. For the second region, ρ/r is regular, and the Laplacian of V is readily expanded in spherical coordinates to obtain

$$\frac{1}{4\pi \epsilon} \int \frac{1}{r} \frac{\partial^2}{\partial r^2} \rho \left(t - \frac{r}{v}\right) dv$$

Since both regions contribute to the potential at the point in question, the complete expression to be substituted for the Laplacian in equation 376 is

$$\nabla^2 V = -\frac{\rho}{\epsilon} + \frac{1}{4\pi \epsilon} \int \frac{1}{r} \frac{\partial^2}{\partial r^2} \rho \left(t - \frac{r}{v}\right) dv$$

Thereafter the solution proceeds without trouble, and equation 376 reduces to an identity. See, for example, Abraham and Becker's *Classical Electricity and Magnetism*.

RADIATION

Let us consider a short length of wire carrying alternating current:

$$i = I \sin \omega t \qquad [380]$$

The wire is isolated in space (there is no ground surface near). Its length is l, and we will locate a set of spherical coordinates in such a way that the conductor extends along the polar axis from $-l/2$ to $+l/2$; see Fig. 58. Using equation 379, we write the vector potential about this short wire; the integration need be performed only along the wire, in the x direction, from one end of the wire to the other, as follows:

$$A_x = \frac{1}{4\pi} \int_{-l/2}^{l/2} \frac{I \sin \omega \left(t - \frac{r}{v}\right)}{r} dx \qquad [381]$$

Since current is in the x direction only, there is only an x component of vector potential. (This may be compared with equation 219.)

Now, if the length of the wire is small compared to the distance at which **A** is measured, the denominator of the integrand is practically constant during the course of integration. If the length is small compared with the wavelength of the radiated signal, the numerator is also practically constant; this means that at any point in space the phase of a signal from one end

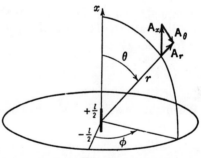

Fig. 58

of the radiating wire is negligibly different from the phase of the signal arriving at the same point from the other end of the wire. With these assumptions the integrand of equation 381 is merely a constant and

$$A_x = \frac{Il}{4\pi r} \sin \omega \left(t - \frac{r}{v}\right) \qquad [382]$$

Thus simply is the vector-potential field found from the known current.

To find the magnetic field about the short antenna, the curl of the vector potential is determined. This may best be done in spherical coordinates, using the formula of Table II. Vector potential is readily changed to spherical components, as illustrated in Fig. 58, giving

$$A_r = \frac{Il}{4\pi\,r} \sin \omega \left(t - \frac{r}{v}\right) \cos \theta$$

$$A_\theta = -\frac{Il}{4\pi\,r} \sin \omega \left(t - \frac{r}{v}\right) \sin \theta \qquad [383]$$

$$A_\phi = 0$$

Taking the curl gives, in accordance with equation 211, the magnetic field:

$$H_r = 0$$

$$H_\theta = 0 \qquad [384]$$

$$H_\phi = \frac{Il}{4\pi\,r} \sin \theta \left[\frac{\omega}{v} \cos \omega \left(t - \frac{r}{v}\right) + \frac{1}{r} \sin \omega \left(t - \frac{r}{v}\right)\right]$$

The electric field can be found from the magnetic field by Maxwell's equation, equation 240, or from the vector potential by equation 371. To illustrate the latter method, we must first find the scalar potential V from equation 374

$$V = -\frac{1}{\epsilon} \nabla \cdot \int \mathbf{A}\, dt \qquad [385]$$

When this is substituted into equation 371 there results an expression entirely in **A**:

$$\mathbf{E} = \frac{1}{\epsilon} \nabla \nabla \cdot \int \mathbf{A}\, dt - \mu \frac{\partial \mathbf{A}}{\partial t} \qquad [386]$$

This somewhat disturbing array of symbols indicates operations that are easily carried out one at a time, giving

$$E_r = \frac{Il}{2\pi\,\epsilon r} \cos \theta \left[\frac{1}{vr} \sin \omega \left(t - \frac{r}{v}\right) - \frac{1}{\omega r^2} \cos \omega \left(t - \frac{r}{v}\right)\right]$$

$$E_\theta = \frac{Il}{4\pi\,\epsilon r} \sin \theta \left[\frac{\omega}{v^2} \cos \omega \left(t - \frac{r}{v}\right) + \frac{1}{rv} \sin \omega \left(t - \frac{r}{v}\right)\right. \qquad [387]$$

$$\left. - \frac{1}{\omega r^2} \cos \omega \left(t - \frac{r}{v}\right)\right]$$

$$E_\phi = 0$$

It is easier to discover the physical meaning of equations 384 and 387 if they are expressed in terms of wavelength λ and frequency f. Using (see Table III):

$$\lambda = \frac{v}{f} = \frac{2\pi v}{\omega} \quad \text{and} \quad \eta = \sqrt{\frac{\mu}{\epsilon}} \quad [388]$$

they may then be written:

$$E_r = -\eta \frac{Il \cos \theta}{r\lambda} \left[\frac{1}{4\pi^2} \frac{\lambda^2}{r^2} \cos\left(2\pi \frac{r}{\lambda} - \omega t\right) + \frac{1}{2\pi} \frac{\lambda}{r} \sin\left(2\pi \frac{r}{\lambda} - \omega t\right) \right]$$

$$E_\theta = \eta \frac{Il \sin \theta}{2r\lambda} \left[-\frac{1}{4\pi^2} \frac{\lambda^2}{r^2} \cos\left(2\pi \frac{r}{\lambda} - \omega t\right) - \frac{1}{2\pi} \frac{\lambda}{r} \sin\left(2\pi \frac{r}{\lambda} - \omega t\right) \right. \quad [389]$$

$$\left. + \cos\left(2\pi \frac{r}{\lambda} - \omega t\right) \right]$$

$$E_\phi = 0$$

$$H_r = 0$$
$$H_\theta = 0$$
$$H_\phi = \frac{Il \sin \theta}{2r\lambda} \left[-\frac{1}{2\pi} \frac{\lambda}{r} \sin\left(2\pi \frac{r}{\lambda} - \omega t\right) + \cos\left(2\pi \frac{r}{\lambda} - \omega t\right) \right] \quad [390]$$

Let us consider these equations in two general regions: first, near the radiating wire, in the region where r is small compared to the wavelength λ, and second, at a distance of several wavelengths so that r is large compared to λ. In the region near the antenna, the terms containing λ/r in the highest degree predominate. Quite close to the antenna, we may disregard all terms except the first in each bracket in equations 389 and 390, and when this is done the equations reduce to the quasi-stationary equations of an oscillating doublet.[3] These terms give what is called the **induction field** about the antenna, that is, the field that neglects radiation.

If, on the other hand, the field is observed at a distance of many wavelengths from the source, so that λ/r is small, another interesting and important simplification appears. In this case terms containing λ/r and λ^2/r^2 are so small that they may be neglected. Only the last term need be retained in the expression for E_θ, and the entire expression for E_r is negligible compared to E_θ. Also, only the last term for H_ϕ is significant. With these approximations, which are good at a distance

---

[3] See equation 220 and Problem 11, Chapter VI.

of several wavelengths from the origin, the wave equations describe what is called the **radiation field**:

$$E_r = 0$$

$$E_\theta = \eta \frac{Il \sin \theta}{2r\lambda} \cos\left(2\pi \frac{r}{\lambda} - \omega t\right) \quad [391]$$

$$E_\phi = 0$$

$$H_r = 0$$

$$H_\theta = 0 \quad [392]$$

$$H_\phi = \frac{Il \sin \theta}{2r\lambda} \cos\left(2\pi \frac{r}{\lambda} - \omega t\right)$$

These equations 391 and 392 describe a beautifully simple electromagnetic field. It is a wave traveling radially outward. The electric and magnetic components are identical in form and are mutually perpendicular. Their magnitudes are related by

$$E_\theta = \eta H_\phi \quad [393]$$

(a relation similar to that found for plane waves in equation 271). The electric and magnetic components become weaker as the wave travels outward because both are inversely proportional to the radius. The Poynting vector that describes the flow of energy is radially outward and is inversely proportional to the square of the radius; this shows that there is no loss of energy and that the energy density merely diminishes as the wave spreads.

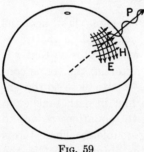

Fig. 59

Figure 59 shows the appearance of one section of the wave. It is a spherical wave. Lines of the magnetic field are parallels of latitude on the sphere, and the electric field is along meridians. Both fields are strongest near the equator and vanish at the poles. The fields at any fixed point in space are, of course, oscillating sinusoidally.

Any small portion of the spherical traveling wave cannot be distinguished from a plane wave. The similarity of Fig. 59 to Fig. 48a is obvious. In equations 391 and 392, the term $\cos(2\pi r/\lambda - \omega t)$ may be considered to define a plane wave, and the coefficient of that term is interpreted as giving the strength of the wave in different parts of space. This is an approximation based upon the fact that, if only a small portion of the wave is observed, neither sin θ in the numerator nor r in the denominator of the coefficient can change appreciably in the region under observation. Derivatives of the coefficient will therefore be vanishingly small. It is for this reason that a received radio wave can usually be considered a plane wave.

The Spherical Wave. There is a great deal of information in equations 389 and 390, the complete expressions for the spherical wave, that

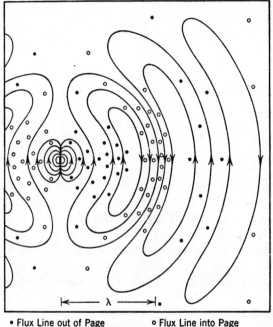

• Flux Line out of Page   ○ Flux Line into Page
Fig. 60

is lost in the approximate expressions 391 and 392. (It should be remarked in passing that the latter are not precisely solutions of Maxwell's equations, and it is not to be expected that they would be.) Figure 60 gives a general idea of the fields described by equations 389 and 390; a cross section is shown in a plane containing the antenna, with lines indicating electric field and small circles or dots to show where magnetic flux enters or issues from the plane. The magnetic flux lines are circular, about an axis containing the antenna. There is actually a much greater concentration of field very near the antenna than can be shown in the diagram.

It is indicated in Fig. 60 that electric lines in the radiated wave do not terminate on charge but are closed curves. The electric field has no divergence; the curves are closed in the polar regions by the radial component E_r. The radial component vanishes in the equatorial region, while E_θ vanishes at the poles. In the "induction field," close to the antenna, lines terminate on charge on the antenna.

The diagram is drawn for time $t = 0$. As time passes, all the outer part of the diagram expands with the speed of wave propagation, while the inner part, near the wire, pulsates with the current in the antenna.

The phase relations are complicated. Near the wire the electric and magnetic fields are out of phase in time. The magnetic field near the antenna is in phase with the antenna current, whereas the electric field is in phase with the charge on either end of the antenna. These induction components of the electric and magnetic fields contain a relatively large amount of energy that is alternating between the electric and the magnetic fields.

But the induction component is negligibly small at a distance of several wavelengths from the source, for it diminishes with the square or cube of the radius, while the radiation component that diminishes only as the first power of the radius becomes predominant. The radiation component represents energy that is traveling outward and that never returns to the circuit from which it was sent. The radiation components of electric and magnetic fields are in phase with each other. The induction and radiation components of the electric field, in the region where both exist, are in phase opposition, while those of the magnetic field are in phase quadrature; this unexpected result may be explained by considering that the radiation component does not originate directly from current and charge in the antenna but rather from the changes of the induction fields surrounding the antenna.

The induction and radiation terms of both the electric and magnetic fields are equal in magnitude when $\lambda/r = 2\pi$, or at a radius from the origin of approximately one-sixth of a wavelength. Beyond that radius the radiation component becomes predominant in proportion to the distance. Hence it may be concluded that phenomena taking place at distances less than one-sixth of a wavelength from a short antenna are predominantly inductive and that those at greater distances are predominantly the result of radiation. Thus many of the early demonstrations of "wireless telegraphy" were primarily the result of induction.

The distinction between the induction and the radiation terms of equations 389 and 390 is a mathematical one: terms containing certain powers of r are induction, those with other powers of r are radiation. But another distinction is more open to physical interpretation, as follows.

The dynamic state differs from the quasi-stationary state because it takes into account the ability of a changing electric field to induce a magnetic field and of a changing magnetic field to induce an electric field. Radiation is the result. The radiation components of the electric and magnetic fields have no such close relation to charge and current as have the induction fields. They are cut adrift from their source. The electric field of a wave results not from the near-by presence of charge but from a changing magnetic component in the wave; the magnetic field results not from an actual flow of current but from a changing electric field. These are shown in the outer regions of Fig. 60. A wave, of course, could not have originated if there had not somewhere been a charge and a current but, having once been produced, a wave may travel any distance and propagate itself for an unlimited time. Consider, for example, the light waves from an extra-galactic nova that reach us millions of years after the brilliance that created them has ceased and the star become extinct.

PROBLEMS

1. Derive equations 387 for **E** from equations 384 using Maxwell's equation 240.

2. At what distance, in wavelengths, is the radiation component of magnetic field twice the induction component? At what distance is it 100 times?

3. Show that the quasi-stationary terms of equation 387 result from a solution for electric potential about an oscillating doublet in the region in which equation 144 can be used. (Note: The solution is greatly simplified by making use of the fact that the length of the doublet, which may be called l, is small compared to the distance r from the midpoint of the doublet to the point at which the potential and electric field are computed. Thus $r^2 - l^2$ is approximately r^2. Also it is permissible to let $r_1 = r - (l/2) \cos \theta$ and $r_2 = r + (l/2) \cos \theta$.)

4. Show that equation 376 has also an *advanced potential* solution $V = \dfrac{1}{4\pi \epsilon} \int \dfrac{\rho(t + r/v)}{r} dv$ as well as the retarded potential solution of equation 378. What is the physical meaning of this "advanced potential"?

5. A spherical condenser consisting of a metal ball surrounded by a concentric metal shell is discharged by making an electrical connection between the inner and outer spheres. The discharge is oscillatory. Is there radiation? Explain.

6. Find the Poynting vector field of a radiated wave from equations 391 and 392.

CHAPTER XII

Antennas

The previous chapter treated radiation from a short conductor with the assumption that the current was the same throughout the whole length of the conductor. This, of course, is physically impossible unless the short conductor is part of a circuit. It may be part of a longer antenna wire, for instance, and, if so, there will be radiation from each part of the longer wire. Total radiation from the antenna is then found by summation, or integration, of the components of radiation from the many short sections of the antenna.

Short Antennas. Consider a wire of reasonable length but still much shorter than the wavelength of the radiated signal. Consider that this wire is isolated in space, with no ground surface or other disturbing body near by; however, at the middle of the wire there is an oscillator or some source of energy to produce current in the wire. See Fig. 61a. Current flows because of the distributed capacitance of the wire; the current is charging current to the capacitance and tapers in amount from a maximum at the middle of the wire to zero at either end.

As a first approximation, the capacitance of the wire is uniformly distributed (as along a transmission line), making the current proportional to distance from the end of the wire. Current is plotted in this way in Fig. 61b.

To find total radiation, consider the antenna of Fig. 61 made up of many sections, each so short that the current is substantially constant through the length of the section. We wish to find the radiation field at a point several wavelengths distant from the antenna and at a distance, therefore, of many times the length of the antenna. Each section of the antenna will contribute to the electric field at this point according to equation 391. This equation is rewritten below, and, since antennas ordinarily radiate into free space or, what is much the same, into air, c is used instead of v:

$$E_\theta = \frac{\eta I l \sin \theta}{2r\lambda} \cos \omega \left(t - \frac{r}{c}\right) \qquad [394]$$

SHORT ANTENNAS   173

The total field strength at the point of observation is simply a summation of the amplitudes of the components received from each short section of the antenna. The only factor in equation 394 that varies appreciably from one point of the antenna to another is the current I. Each component of radiation is directly proportional to the amount of current in the section of the antenna from which it is radiated. Thus the radiation received from a section near the end of the antenna of Fig. 61 is much less than the radiation from a section of equal length near the middle. Taking into account the distribution of current, the total ra-

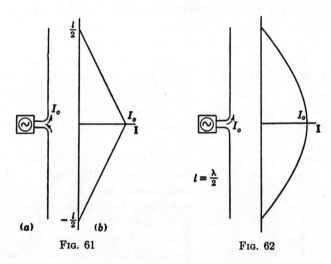

Fig. 61   Fig. 62

diation received at any point from the antenna of Fig. 61 is equal to the radiation that would be received from an antenna half as long if it were possible to have the current in all the antenna equal to the current I_0 at the center.

Such an antenna as that of Fig. 61—isolated, straight, and short compared to λ—is therefore said to have an equivalent length of half its actual length. To compute radiation from the actual antenna, equation 394 is used, but the value substituted for l is the **equivalent** or **effective** length, half the actual length of the antenna.

Practically, the effective or root-mean-square value of the field at the receiving antenna is usually the quantity desired. This may be written in terms of effective length l_e and the effective or root-mean-square current at the midpoint of the antenna I_0 as

$$\text{Effective (rms) field strength in volts per meter} = \eta \frac{I_0 \, l_e}{2r \, \lambda} \sin \theta \quad [395]$$

η for free space is 377 ohms.

I_0 is effective midpoint current in amperes (rms).

l is actual length of the isolated, straight antenna in meters.

l_e is effective length in meters (if $l \ll \lambda$, $l_e = 0.5l$).

θ is angle between transmitting antenna and a line connecting the transmitting antenna to the receiving antenna.

r is distance from transmitting antenna to receiving antenna in meters.

λ is wavelength of signal in meters.

Half-Wavelength Antenna. When the length of an antenna is considerable compared with the wavelength of the signal, treatment as if it were a short antenna is inexact. Two refinements are necessary.

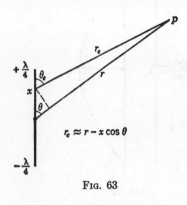

Fig. 63

First, if the antenna length is comparable to the wavelength, current in the antenna will not be proportional to distance from the end of the antenna. An antenna acts much as an open-circuited transmission line with distributed capacitance; if short, the current distribution is substantially linear, as in Fig. 61, but, if the antenna is longer, this approximation is unsatisfactory. Assuming uniform distributed capacitance along the antenna (an assumption that is not precise but reasonably good and quite generally accepted), the current is proportional to the sine of the distance from the end of the antenna. As an example, Fig. 62 shows current distribution in a line that is a half wavelength long; taking current at the center of the antenna as I_0, the amplitude of current I at any distance x from the center of the antenna is

$$I = I_0 \cos 2\pi \frac{x}{\lambda} \qquad [396]$$

A receiving antenna at some distant point will receive a component of signal from each short section of the antenna of Fig. 62, and the strength of each component will be proportional to the value of I in the corresponding antenna section.

A further complication results because the different components of signal may not be in phase with each other when they arrive at the receiving antenna. Consider the received signal at such a point as p in Fig. 63. The distance r_e in the figure is less than r, so radiation travel-

HALF-WAVELENGTH ANTENNA

ing the path of length r would be more delayed in reaching p and would consequently lag in phase behind the radiation traveling the path r_e.

These considerations are to be introduced into equation 394. A short section of antenna of differential length dx, located at a distance x from the midpoint, has an amplitude of current given by equation 396. The distance of this section of antenna from the point of observation p is, as in Fig. 63, r_e. The total field received at p from the antenna is found by integrating equation 394 along the whole length of the antenna, from $x = -\lambda/4$ to $x = +\lambda/4$:

$$E_\theta = \int_{-\lambda/4}^{\lambda/4} \frac{\eta I_0 \sin \theta_e}{2 r_e \lambda} \cos\left(2\pi \frac{x}{\lambda}\right) \cos \omega \left(t - \frac{r_e}{c}\right) dx \qquad [397]$$

Here, as in Fig. 63, r_e is the distance from the current-carrying element to the point at which field strength is being measured; r is the distance from the origin of coordinates at the midpoint of the antenna. Since we are interested in the electric field at distances from the antenna greater than several wavelengths, the difference between r_e and r in the denominator of the integrand is negligible. But in the cosine term the distinction between r_e and r is essential, for it is this difference that determines the phase relation of radiation from different parts of the antenna; $r - r_e$ may not be negligible compared to λ, although it is certainly negligible compared to r. In the denominator it is quite satisfactory to substitute r for r_e, but in the phase relation it is necessary to use an approximation that cannot be in error by more than a small fraction of a wavelength. Referring again to Fig. 63, we see that a very good approximation is

$$r_e = r - x \cos \theta \qquad [398]$$

Finally, the angle θ is not appreciably different from the angle θ_e. With these changes

$$E_\theta = \frac{\eta I_0 \sin \theta}{2 r \lambda} \int_{-\lambda/4}^{\lambda/4} \cos\left(2\pi \frac{x}{\lambda}\right) \cos \omega \left(t - \frac{r}{c} + \frac{x \cos \theta}{c}\right) dx \qquad [399]$$

Performance of this integration, although somewhat involved, is essentially simple. The result is

$$E_\theta = \frac{\eta I_0}{2\pi r} \cos \omega \left(t - \frac{r}{c}\right) \frac{\cos\left(\frac{\pi}{2} \cos \theta\right)}{\sin \theta} \qquad [400]$$

The other components of the electric field are of course zero, as they are for a short antenna.

The magnetic field is perpendicular to the electric field (as it is for each elementary length of antenna) and is related by the intrinsic impedance, giving:

$$H_\phi = \frac{I_0}{2\pi r} \cos \omega \left(t - \frac{r}{c}\right) \frac{\cos\left(\frac{\pi}{2}\cos\theta\right)}{\sin\theta} \qquad [401]$$

The other components of magnetic field are zero. In equations 400 and 401, symbols and units are the same as in equation 395, and H_ϕ is in amperes (or ampere-turns) per meter.

Equations 400 and 401 are only for antennas that are one-half wavelength long. The method used to derive these equations can readily be extended to antennas of any length. A more general expression for current distribution is then used instead of equation 396. However, from both the practical and theoretical points of view, the half-wave antenna is a very interesting example of a long antenna and is the only one that will be worked out in detail here.

It is interesting to compare equation 400 for a half-wave dipole antenna with equation 395 for a short antenna. The effective (rms) field strength at any point in a plane normal to the antenna (for which $\theta = 90$ degrees) is, for a half-wave dipole antenna, $\eta \dfrac{I_0}{2\pi r}$. For a short antenna of effective length l_e it is $\eta \dfrac{I_0}{2r} \dfrac{l_e}{\lambda}$. These formulas give the same result if $l_e = \lambda/\pi$, and for this reason it is said that the effective length of a half-wave dipole antenna is λ/π. Since the actual length of such antenna is $\lambda/2$, the effective length is $2/\pi$ times the actual length.

This tells us that the effective length of a half-wave antenna is $2/\pi$ or 0.637 times its actual length, and we know that the effective length of a very short antenna is 0.5 times its actual length. One cannot, therefore, go radically wrong in estimating the effective length of any dipole antenna shorter than a half wavelength to be about five- or six-tenths of the actual length.

Radiation Pattern. Effective length gives comparison of field strength in a normal direction only. In comparing the distribution of radiation from a half-wave dipole with that from a short antenna in other directions, the "radiation pattern" is useful. Vectors are plotted radially from a point (see Fig. 64), the length of each vector being proportional to the field strength at a given distance from the antenna in the direction indicated by the vector. A curve connecting the ends of the vectors is then the radiation pattern. The field strength is frequently plotted in microvolts per meter at a distance of 1 mile.

For either a short antenna or a half-wave dipole, radiation normal to the antenna is equal in all directions; the radiation pattern in a normal plane is therefore a circle, as in Fig. 64a.

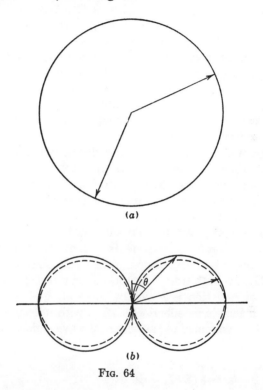

Fig. 64

The radiation pattern of a short antenna in a plane containing the antenna is shown by the solid line in Fig. 64b. The maximum radiation is normal to the antenna, and radiation in the direction of the antenna axis is zero. Radiation in other directions is proportional to the sine of the angle from the axis, as in equation 395, so the pattern is composed of a pair of circles as shown.

It is sometimes useful to think of Figs. 64a and b as being cross sections of a solid surface something like a doughnut. Such a surface is the complete three-dimensional radiation pattern.

The radiation pattern of a half-wave dipole antenna is surprisingly similar. The peculiar trigonometric function in the numerator of equation 400 is numerically similar to $\sin^2 \theta$; this function divided by $\sin \theta$

178 ANTENNAS

is therefore not greatly different from sin θ, the corresponding term in equation 394. The radiation pattern of a half-wave dipole antenna is shown as a dash line in Fig. 64, the scale being adjusted to make radiation normal to the antenna equal the normal radiation from the short antenna.

Since the half-wave dipole radiation pattern of Fig. 64 lies within the short-antenna radiation pattern, it follows that less energy need be radiated from a half-wave dipole antenna than from a short antenna to give the same field strength in the direction normal to the antenna. It is sometimes interesting to plot the two radiation patterns assuming equal radiated energy rather than equal normal field strength. Normal radiation from the half-wave antenna is some 6 per cent greater than from a short antenna radiating equal power. Thus a very small degree of directivity is obtained.

Radiated Power. The total power radiated from an antenna is often of interest. There are several ways in which radiated power can be computed, but one way or another they involve the Poynting vector. As a basic and yet simple example, power from the short antenna of equation 394 will be determined.

Energy is carried away from an antenna by the radiation field, and not by the induction field components. Considering only the radiation component, the Poynting vector is radial and, since it is in the direction of $\mathbf{E} \times \mathbf{H}$, it is outward. The total energy transported by the traveling wave of radiation is found by integrating the Poynting vector over an imaginary large spherical surface with its center at the origin. Because of the symmetry of the spherical wave, this is an easy integration and gives

$$\int \mathbf{P} \cdot d\mathbf{a} = \int (\mathbf{E} \times \mathbf{H}) \cdot d\mathbf{a}$$

$$= \int \eta \left[\frac{Il \sin \theta}{2r\lambda} \cos \omega \left(t - \frac{r}{c} \right) \right]^2 (2\pi r^2 \sin \theta) \, d\theta$$

$$= \frac{\eta \pi I^2 l^2 \cos^2 \omega \left(t - \dfrac{r}{c} \right)}{2\lambda^2} \int_0^\pi \sin^3 \theta \, d\theta$$

$$= \frac{2\eta \pi I^2 l^2}{3\lambda^2} \cos^2 \omega \left(t - \frac{r}{c} \right) \qquad [402]$$

The Poynting vector is a function of time, varying as the square of the cosine. For most purposes, the average power of the radiated wave is desired, and, since the average value of the cosine squared function is ½, it follows that the average power radiated from a short isolated antenna with uniform current distribution is

$$\text{Average power} = \frac{\eta \pi I^2 l^2}{3\lambda^2} \quad [403]$$

Radiation Resistance. A term that is defined as the average radiated power divided by the square of the effective value of current in the antenna lead is called **radiation resistance**. In equation 403, I is the *maximum* value of current; the square of the *effective* current is $\frac{1}{2}I^2$, so the radiation resistance of a short antenna is

$$\frac{2\pi \eta l^2}{3\lambda^2} \quad [404]$$

This formula is derived for a theoretical doublet and can be applied to a short dipole antenna if the *equivalent* length l_e is used for l. Introducing the value of η for free space, radiation resistance of a short dipole antenna of equivalent length l_e is

$$789 \left(\frac{l_e}{\lambda}\right)^2 \text{ohms} \quad [405]$$

This is not correct for a half-wave dipole, however, or for any antenna with a radiation pattern appreciably different from that of a doublet, even though equivalent length is used.

To find the radiation resistance of a half-wave dipole, or any other antenna, the appropriate functions for **E** and **H** are substituted in equation 402 and integrated. For the half-wave dipole antenna, **E** and **H** from equations 400 and 401 would be used. The result is simple, although the computation is somewhat involved. *Radiation resistance of a half-wave dipole antenna, for any frequency, is 73.1 ohms.*

Antennas Above Ground. Practically, antennas are not isolated in space; the surface of the earth is rarely very far away. Antennas for operation at standard broadcast or lower frequencies are usually vertical and take advantage of the presence of ground to double the effective length of the antenna. High-frequency antennas, on the other hand, are commonly horizontal dipoles at some distance above ground, and the signal received at any distant point will consist of both ground-reflected radiation and radiation received directly. The ground-reflected radiation may be advantageous or disadvantageous, depending on the dimensions.

Consider first a horizontal antenna at a distance h_1 above ground, as in Fig. 65, with a receiving antenna that is h_2 above ground. At the receiving antenna there is a total field comprising both radiation received directly and radiation received after reflection from ground. Radiation is horizontally polarized (the diagrammatic representation of field strength in the figure may, if one wishes, be taken to represent the magnetic component of the wave).

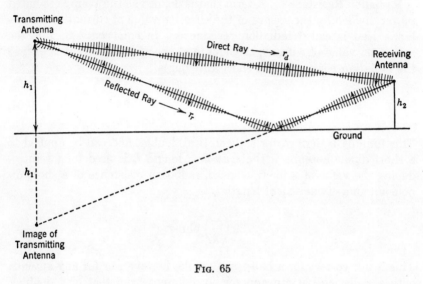

FIG. 65

Because reflected radiation travels a longer path, and because of phase shift resulting from reflection, the two components of field at the antenna will not, in general, add in phase with each other.

In Chapter X attention was given to oblique reflection from the surface of an imperfect insulator such as the earth. A general expression for the reflection coefficient for a horizontally polarized wave was obtained. It was shown that the reflection coefficient for a horizontally polarized wave reflected from ordinary earth is not greatly different from -1, and that when the wave approaches the earth's surface at nearly glancing incidence the value of -1 is almost exactly correct. This means that, if the antenna heights are small compared to the distance, the reflected ray of Fig. 65 will be reversed in phase by reflection, but its amplitude will not be diminished.

If the length of path followed by the reflected ray were the same as that of the direct ray, the two components would be exactly out of phase at the receiving antenna and would completely cancel. No signal would be received. This unfortunate situation would be approached if either h_1 or h_2 were very small. But, if the antennas are high enough to make the length of path of the reflected ray just one-half wavelength longer than the path of the direct ray, the two will add and give a doubly strong signal. For other heights, the fields of the two rays will add to give a resultant that is determined by the difference of path lengths.

From here on, the problem is one of geometry. If the length of path of the direct ray is r_d, and that of the reflected ray r_r, and the horizontal distance is d, and if curvature of the earth is neglected:

$$r_d = \sqrt{d^2 + (h_1 - h_2)^2}$$
$$r_r = \sqrt{d^2 + (h_1 + h_2)^2}$$
[406]

Expanding each by the binomial series, retaining only two terms of each, and subtracting, gives the difference of path length which may be called Δr:

$$\Delta r = r_r - r_d \approx \frac{2h_1h_2}{d} \quad [407]$$

Signal strength from the direct ray is given by

$$E_d = \frac{\eta I l \sin \theta}{2r_d \lambda} \cos \omega \left(t - \frac{r_d}{c} \right) \quad [408]$$

Considering the phase reversal after reflection, the reflected ray is

$$E_r = -\frac{\eta I l \sin \theta}{2r_r \lambda} \cos \omega \left(t - \frac{r_r}{c} \right) \quad [409]$$

The difference between r_d and r_r in the coefficient is negligible, and adding the direct and reflected components gives

$$E = \frac{\eta I l \sin \theta}{2r\lambda} \left[\cos \omega \left(t - \frac{r_d}{c} \right) - \cos \omega \left(t - \frac{r_r}{c} \right) \right] \quad [410]$$

A trigonometric change introduces no further approximations:

$$E = \frac{\eta I l \sin \theta}{2r\lambda} \left[2 \sin \omega \left(\frac{\Delta r}{2c} \right) \sin \omega \left(t - \frac{r_d}{c} - \frac{\Delta r}{2c} \right) \right] \quad [411]$$

Since $\omega/c = 2\pi/\lambda$, the *magnitude* of the total received field can be expressed in terms of the field of the direct ray alone as

$$|E| = |E_d| 2 \sin \frac{2\pi h_1 h_2}{\lambda d} \quad [412]$$

It is clear that this approximation applies under rather restricted conditions, but they happen to be the practical conditions for many cases of ultra-high frequency transmission, and this expression is widely used.[1]

Figure 65 suggests the use of an image of the transmitting antenna to replace the effect of ground reflection. This assumes that the ground is a perfect reflecting surface. The geometry of the image antenna is exactly what an observer would see looking into the perfect reflecting plane as a mirror (and this is only natural when it is realized that a mirror *is* a reflecting plane, and the general principle that the mirror produces an image is equally valid for light waves or radio waves). But, because of phase reversal, the electric current and charge that must be assumed in the image are exactly opposite to the current and charge the observer would see (if current and charge were visible) as he looked in the mirror. This is a general rule for antennas of any configuration above a perfect reflecting surface. It is evident in Fig. 65, and it can be extended to more complicated arrangements by consideration of reflected rays.

Practically, it is undesirable to have horizontal antennas too low. Equation 412 gives the amplitude of the total field at a receiving antenna in terms of the field E_d that would exist if there were no reflection. The strength of E_d, the direct ray, varies inversely as distance. For relatively small antenna heights, the total field E will vary inversely as the *square* of distance. Therefore, if antenna heights and other factors remain the same, received energy varies inversely as the *fourth power* of distance. This is true if $h_1 h_2$ is much smaller than $4\lambda d$, clearly an unfavorable arrangement.

The range of radar for detecting low targets is strongly influenced by this effect of cancellation of signal resulting from ground reflection. If energy reradiated by a target is proportional to energy impinging on the target, and if energy at the target varies as the fourth power of distance from the transmitter, the reflected energy received back at the point of transmission is inversely proportional to the *eighth power* of the target distance. (This is a first approximation for a low target, a distance of several kilometers, and a wavelength of decimeters or more, and it assumes a fairly smooth ground or sea surface from which radiation is specular and not merely diffuse.)

Grounded Antennas. Vertical antennas, with voltage applied between the base of the antenna and ground, are quite commonly used

---

[1] See, for this use, and for many other antenna applications, *Radio Engineers' Handbook*, F. E. Terman, McGraw-Hill Book Co., New York, 1943, Sections 10 and 11.

for frequencies less than a few megacycles. Radiation received at any distant point, as P in Fig. 66, will comprise a direct and a reflected component. If the reflection from ground is perfect, with complete phase reversal, we may consider the ground-reflected ray to come from an image antenna, as discussed above. The problem of radiation from an antenna-and-its-image is simpler than the problem of an antenna-and-a-reflecting-plane.

Reflection from the surface of the earth is not perfect, but it may be quite good for typical practical values of frequency and conductivity. In an earlier chapter, the reflection of horizontally polarized waves was

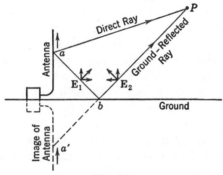

Fig. 66

considered. Radiation from a vertical antenna will be vertically polarized (or, more accurately, polarized in a vertical plane containing the ray), and its reflection characteristics are different, as follows.

If the earth were a perfect conductor, the reflection coefficient would be -1 for all angles of incidence, because the sum of the incident and reflected tangential components of electric field must always be zero. That is, the *horizontal component* of the electric field (see Fig. 66) would be completely reversed by reflection. The vertical component would not be reversed; it will be seen from the diagram that an unchanged vertical component is necessary to accompany a reversed horizontal component after reflection.

If, on the other hand, the earth were a perfect dielectric material, the reflected wave would be smaller than the incident wave, the magnitude of the reflection coefficient being greatly dependent upon the angle of incidence. Indeed, at one particular angle, known as the Brewster angle, all energy falling on the surface would enter the dielectric, and reflection would be zero. Whether or not there is phase reversal of the reflected ray depends on whether the angle of incidence is greater or less than this Brewster angle. Comparable to equation 364, it is not difficult to derive for such a wave, polarized in what is called the plane of incidence, the following reflection coefficient:

$$\frac{E_{t2}}{E_{t1}} = \frac{\eta_3 \sin \psi_3 - \eta_1 \sin \psi_1}{\eta_3 \sin \psi_3 + \eta_1 \sin \psi_1} \qquad [413]$$

E_{t1} and E_{t2} are the tangential components of the incident and reflected waves, respectively; η_1 and η_3 are the intrinsic impedances in the media of the incident and refracted waves; and ψ_1 and ψ_3 are the angles between the incident and refracted rays, respectively, and the surface of reflection (it should be noted again that the conventional angles of incidence and refraction are the complements of these). Equation 413 is applicable to either a dielectric or a poor conducting medium.

Clearly it is easier to consider the earth a perfect conductor than a poor conductor, and whether this may reasonably be done depends largely on frequency. If conduction current in the earth is much greater than displacement current, the earth will reflect as a good conductor. Let us determine the frequency for which conduction current is ten times displacement current; that is, for which $\gamma/\omega\epsilon = 10$. This will indicate the order of magnitude of the highest frequency for which the earth is a good conductor. Table V gives some typical values of

TABLE V

EARTH CHARACTERISTICS

	Relative Dielectric Constant, κ	Resistivity in Meter-Ohms, $1/\gamma$
Sea water	81	0.20–0.25
Fresh water	81	50–1000
Fresh-water marsh	15–25	10– 100
Moist, rich, pastoral soil	12–15	50– 250
Dry, rocky, sandy soil	10–14	500–1000
Solid rock	5–10	1000 and up

earth characteristics. Using these, we find that sea water fails to reflect as a good conductor above about 90 megacycles; typical moist earth at a little over 1 megacycle; dry sandy soil above 200 kilocycles. These values give a rough indication of the frequency range to which the following discussion, based on the assumption of a perfectly conducting ground, may be expected to apply. At higher frequencies the assumption of perfect reflection may or may not introduce serious error, depending on the dielectric constant, the angle of incidence, and (most important) the use to be made of the data.

If the antenna is a single short vertical wire above a perfect ground, it gives a radiation pattern above the ground surface that is identical with that from a dipole antenna consisting of the actual antenna and its image. Equation 395 therefore applies, and, since the effective height h_e of the antenna above ground is half the effective length of the antenna-plus-image combination, that equation becomes

$$\text{Effective (rms) field strength in volts per meter} = \eta \frac{I_0 h_e}{r \lambda} \sin \theta \qquad [414]$$

I_0 is rms current at the base of the antenna, and other symbols are as in equation 395.

Equation 414 is intended for use with an antenna that is short compared to the wavelength. If h/λ, the height in wavelengths, is less than $\frac{1}{8}$, the effective height of a vertical wire is practically half the actual height.

It would clearly be advantageous to increase the effective height of an antenna. This is done by increasing capacitance at the top of the antenna, and thereby both increasing the total current and improving the current distribution. Enough capacitance at the top will make current in the antenna practically uniform, and the effective height is then equal to the actual height. Even if current at the base of the antenna were not increased, this would double the radiation field strength and multiply the energy radiated by four.

Any large metal surface at the top of the antenna will provide the necessary additional capacitance. One very common method is to use a "flat-top" antenna, connecting the vertical wire to one or more horizontal wires in a T or inverted-L form. The primary purpose of the horizontal wires is to increase capacitance and thereby to increase current in the vertical wire; there is some radiation from the horizontal wires also, but that is incidental and may usually be neglected. Equation 414 is applicable to such an antenna, with h_e equal to the actual height if the total length of horizontal wire is several times the height.

Radiation resistance of an antenna above ground differs from that of a dipole antenna because, although there is the same radiated energy in the hemisphere of space above ground, there is none below the surface. Half as much energy is radiated, and radiation resistance is therefore half as great. Writing $2h_e$ for l_e in equation 405 and dividing by 2 because the energy is half gives the radiation resistance of a short vertical or flat-top antenna of equivalent height h_e above ground:

$$\text{Radiation resistance} = 1578 \left(\frac{h_e}{\lambda}\right)^2 \text{ ohms} \qquad [415]$$

This formula, in fact, is a good approximation for a vertical antenna as much as a quarter wavelength high if the proper equivalent height is used.

The radiation patterns from grounded antennas are, of course, similar to those of dipole antennas of twice the length. Thus the solid lines of Fig. 64 apply to a short vertical antenna above ground, and very nearly to a flat-top antenna. The lower half of Fig. 64b is meaningless for a grounded antenna, however, for it would refer to a region below the ground surface. Similarly, the dash lines of Fig. 64 show the radiation pattern from a straight vertical antenna that is a quarter wavelength high. Up to this height, the radiation pattern is not very different from that of a short antenna.

However, if the length of the antenna above ground is much greater than a quarter wavelength, the radiation pattern is radically altered. Up to about six-tenths of a wavelength, the horizontal radiation is increased, and the radiation pattern becomes a long, low loop rather than a half-circle. Still longer antennas radiate upward at an angle, with a relatively weak ground wave. The proper choice of height (particularly on an economic basis) is very important in antenna design.

Antenna Arrays. When an array of sending antennas is used, it is sometimes possible to gain a good deal of information by superposition of the fields from the component radiating elements. For example, two vertical antennas that are excited in time quadrature and that are located one-quarter wavelength apart, as in Fig. 67, will radiate in the direction from the leading antenna to the lagging antenna, but not in the opposite direction. The reason is that a wave traveling from the leading antenna to the lagging antenna will be reinforced by the wave from the latter, for the quarter-cycle phase lead of the former wave will just compensate for the quarter-cycle lead in space of the latter, while the wave traveling from the lagging to the leading antenna will be a half-cycle out of phase with the wave from the leading antenna, and hence the resultant field strength (at a considerable distance) in that direction will be zero. This is only one of many antenna arrays of practical importance in obtaining desired radiation patterns.[2]

A radiation pattern of particular interest is one that has a concentrated beam like a searchlight, with little energy spreading in other directions. A sharp beam can be achieved by using a very great number of radiating elements. If the radiating elements are so many and

---

[2] A discussion of radio antennas, for which this chapter provides the theoretical basis, will be found in Chapter 14 of *Radio Engineering* (Third edition) by F. E. Terman, McGraw-Hill Book Co., New York, 1947.

so close that they merge into a single conducting surface, they become equivalent to a mirror. Indeed, a single radiator at the focus of an actual mirror can be used to gain directivity with short radio waves just as with even shorter light waves.

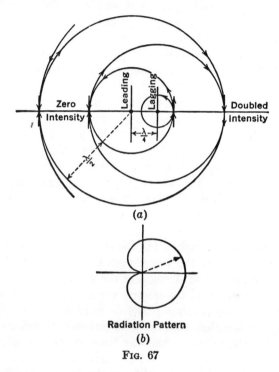

Fig. 67

Receiving Antennas. As an electromagnetic wave travels through space, a changing magnetic field continually produces an electric field, and the resulting electric field continually reproduces a magnetic field. When the wave passes any given point in space, the changing magnetic field induces an electric field at that point, and, if the wave passes a wire of conducting material, it induces an electric field in that wire.

There are two general ways to consider the action of the conductor as a receiving antenna. The most straightforward is to recognize that the passing wave will be distorted by the presence of conducting material; its electric and magnetic fields will be weakened because there can be no field strength within a perfect conductor. Current will flow in the conductor, providing proper termination for the electric and magnetic fields of the wave. From this point of view the receiving antenna is a boundary problem.

The alternative method is indirect, but it is more useful for practical computation. It assumes that electromotive force is induced in the antenna by the magnetic field of the *undistorted* passing wave, that as a result current flows, and that this current acts in the antenna as if it were a transmitting antenna and produces a new electromagnetic field that is superimposed upon the undistorted wave. The flow of current in this case is limited by the distributed capacitance and inductance of the conductor, and by its resistance (loss resistance) and "radiation resistance." The problem thus becomes a circuit problem and is much easier to handle.

We will divide radio receiving antennas into three classes for discussion, although any such classification is quite arbitrary. First we will consider a straight wire that is isolated from ground. Current flowing in such a wire will be charging current that can flow from end to end, limited by the capacitance of one end of the wire relative to the other. The conductor shown in Fig. 45, page 101, is an example, and in the alternating magnetic field of a wave there will be alternating current flowing from end to end of such an antenna at the frequency of the passing wave.

The electromotive force induced in the entire length of a straight wire that is short compared to the wavelength is the product of the length of the wire, the electric field strength of the wave, and the cosine of the angle between the wire and the electric vector. (In a longer wire it is necessary to take into account phase differences.) The *effective* electromotive force is less than this, however, for only an infinitesimal part of the antenna current flows all the way from one end of the wire to the other. The effective electromotive force depends upon the distribution of capacitance and can be increased by increasing the capacitance near the end of the antenna.

The *effective length* of a receiving antenna is the same as the effective length of the same antenna as a transmitter. The effective electromotive force in the receiving antenna is the field strength times the effective length: $V = \mathbf{E} \cdot \mathbf{l}_e$. The effective length of a short, straight antenna is half the actual length.

In making use of an isolated straight wire as an antenna, a radio receiving apparatus is located at its midpoint. As much energy as possible is abstracted from the oscillating antenna current by the receiving apparatus. To increase the antenna current, inductance may be inserted as part of the receiving apparatus, providing resonance with the distributed capacitance of the antenna. The current that flows in such a "tuned" antenna is limited only by the energy lost in resistance and the energy reradiated from the antenna, for inductance and capacitance are balanced against each other. Tuning gives the optimum operation of an antenna. A tuned antenna can be considered as an oscillatory circuit; its voltage, from end to end, may be many times the electromotive force induced by the passing wave, for the induced electromotive force is merely enough to maintain the natural oscillation of the antenna current. An antenna one-half wavelength long is naturally tuned without the addition of inductance, for the distributed inductance of the wire itself just balances its distributed capacitance.

For practical computation it is possible to devise an equivalent circuit, in which an equivalent lumped voltage (representing the effective electromotive force of the antenna) drives current through an equivalent lumped antenna impedance in series with the impedance of the receiving apparatus.[3] Antenna design is based on an equivalent circuit of this kind.

An antenna in the form of a vertical straight wire with one end connected to ground will next be considered. Radio receiving apparatus is introduced into the antenna at the point of connection to ground. This is the most familiar type of receiving antenna. It receives the component of a passing wave that is polarized with the electric field vertical. It is non-directional in a horizontal plane, for the directional characteristics of a receiving antenna are the same as those of the same antenna used for transmitting. This follows from the *Reciprocity Theorem*.[4]

The operation of a grounded antenna is not essentially different from that of an isolated antenna of twice the length. Current is limited by the capacitance from antenna to ground, and the greater capacitance of a grounded antenna compensates for the lower electromotive force induced in its shorter length. The effective value of the induced electromotive force is increased, and the impedance of the antenna to ground is decreased, if there is a relatively large part of the capacitance to ground near the top of the antenna; this is often provided by connecting horizontal wires at the top of a vertical conductor.

A loop antenna is the third form to require discussion. Its operation is not essentially different except that flow of current in the loop is not limited by capacitance but by inductance in series with the receiving apparatus. Electromotive force is induced in the loop in the same way

---

[3] For discussion of this and other engineering applications see *Radio Engineering*, F. E. Terman, Third Edition, McGraw-Hill Book Co., New York, 1947.

[4] The Reciprocity Theorem: the positions of an impedanceless generator and ammeter may be interchanged without affecting the ammeter reading. This was applied to radiation by J. R. Carson and others. See, for instance, *Electromagnetic Waves* S. A. Schelkunoff, D. Van Nostrand Co., New York, 1943, pages 476–479.

that it is in any conductor; it is induced by the changing magnetic field and is equal to the integral of the induced electric field around the loop.[5]

Consider for simplicity a rectangular loop with the sides vertical and the top and bottom horizontal. A passing wave is polarized with the electric vector vertical. Electromotive force will be induced by such a wave in the vertical members of the loop, but there will be none in the horizontal members. If the voltages induced in the two vertical members are identical, there will be no current in the loop. This condition results when the plane of the loop is parallel to the plane of the wave. But, if the loop is turned 90 degrees so that its plane is normal to the plane of the wave and parallel to the direction of wave propagation, the voltages in the two vertical members will be somewhat out of phase. The oncoming wave will reach one side of the loop before it reaches the other. The total electromotive force around the loop will then be the difference of the two induced voltages, and not zero. If the loop is small compared with the wavelength of the received signal, the induced electromotive force in either vertical member is proportional to the height of the loop; the phase difference between voltages in the vertical members is proportional to the width of the loop; the electromotive

---

[5] A question that very commonly arises in reference to receiving antennas is: Is the antenna voltage produced by the electric field of the passing wave, or the magnetic field, or both? This is a natural question, but the answer is clear when it is considered that *anywhere in space* the electric field of a traveling wave is the result of a changing magnetic field. The electric field induced in an antenna is likewise the result of the changing magnetic field, and whether one wishes to consider the electromotive force as the integral of the electric field of the wave in space (which it is) or as produced by the change of magnetic field (which it also is) is immaterial. The above question is analogous to asking whether a cork rising on the crest of a water wave is lifted by increasing pressure or by the higher water level; in wave motion there cannot be one without the other.

If, however, a receiving antenna is close to an electric disturbance of some kind, conditions are quite different. Near the source of the disturbance the induction fields predominate, and the radiation fields are negligible. The electric induction field emanates from near-by charge, as distinguished from the radiation field that is produced by changing magnetic field. A metal shield will protect a loop antenna from the *electric induction field* of near-by disturbances, for, like an electrostatic field, the induction field will not penetrate a closed metal surface. But a shield (in the usual form of a pipe that contains the wires of the loop) will not appreciably decrease the amount of magnetic flux that passes through a loop and links with it when a wave is going by. If the shield could act as a short-circuited turn, it would reduce the magnetic field linking with the loop, but a shield for a loop antenna is designed with an insulating section so that it will not carry current around the loop. Hence a radio signal is received on an antenna inside the shield, although much of the noise brought by induction from near-by disturbances is eliminated. Antennas other than loops cannot be shielded, for their operation depends on their capacitance to ground, which a shield would eliminate.

force around the entire loop is proportional to the product of height and width, and therefore to the area of the loop.

The same conclusion is reached from a slightly different point of view, although the essential principle is the same, if induced electromotive force in the loop is considered proportional to the rate of change of magnetic flux linkages through the loop (equation 191). It is then immediately evident that the area of the loop is a controlling factor and that the loop will receive a maximum signal when its plane is normal to the magnetic field of the wave.

Loop antennas commonly have more than one turn of the antenna wire, and the antenna voltage is proportional to the number of turns. It is quite simple to compute the voltage between the terminals of a loop antenna if no current is allowed to flow. But when current flows the distributed inductance is important; the resistance of the wire and reradiation from the loop as well as the impedance of the receiving apparatus must be considered, and these quantities, together with the antenna inductance, are either computed or measured.

PROBLEMS

1. Compare the equations of this chapter with expressions for radiation field strength in another book, such as equation 14–1, Chapter 14, of *Radio Engineering* (3rd edition) by F. E. Terman or equation 35, Chapter XIX, of *Communication Engineering* by W. L. Everitt.

2. An isolated center-fed antenna 2 meters long is radiating a signal with 20-meter wavelength. The current at the antenna midpoint is 100 milliamperes. What is the radiated field strength at a distance of 500 meters? at 1 mile? Use the direction of maximum field.

3. Integrate equation 399 and obtain equation 400.

4. An isolated antenna is one wavelength long. Compute an expression for radiation field similar to equation 400. Plot the radiation pattern. Can you express an "effective length"?

5. Change equation 395 to be in terms of antenna voltage and capacitance (end to end) instead of current. If voltage is held constant as frequency is changed, to what power of frequency is radiated energy proportional?

6. In Problem 2, what is the radiated power per square meter at a distance of 500 meters?

7. What is the radiation resistance of the antenna of Problem 2? What power is being radiated?

8. An isolated antenna one-half wavelength long is excited at its midpoint. What determines the voltage that must be applied to maintain the required antenna current?

9. A horizontal half-wave dipole antenna is 30 meters above ground. It is transmitting a 1000-megacycle signal. The receiving antenna is 10 miles away and 10 meters high. Assuming perfect reflection from a plane earth, what fraction of the direct ray is the total received signal strength?

10. The antennas of Problem 9 are to be raised or lowered to provide maximum received signal strength. Determine heights to be used. Discuss practicability of such heights.

11. What is the radiation resistance of a straight vertical antenna one-fourth wavelength high, rising from a perfect ground plane? The antenna is excited at the base.

12. A straight vertical antenna is 0.6 wavelength high, rising from a perfect ground plane. Compute an expression for radiation field similar to equation 400. Plot the radiation pattern. What is the "effective height"? (Note: This height gives the optimum ground wave.)

13. Use equation 415 to find the radiation resistance of a simple vertical antenna with a height of one-fourth wavelength above ground. Use an effective height of 0.6 times the actual height. Find the radiation resistance of a similar antenna if the radiated wavelength is 5.6 times the height. Use an effective height of 0.55 times the actual height. (See Ballantine "On the Radiation Resistance of a Simple Vertical Antenna," *Proc. Inst. Radio Eng.*, Volume 12, 1924, pages 823–832.)

14. For the antennas of Problem 9, what is the reflection coefficient if the earth is good farm land? Would an assumption of perfect reflection be justified?

15. Repeat Problem 14 for a vertically polarized signal. Is there reversal of phase of the vertical component of electric field at reflection? Would an assumption of perfect reflection be justified?

16. A wave in air, $\lambda = 3.0$ centimeters, falls on glass, $\kappa = 4.00$. Neglect loss. Find the Brewster angle.

17. A loop antenna 1 meter square, with ten turns of wire, is used to receive a radio signal. The signal is transmitted from a vertical antenna that rises to a height of 5 meters above ground. The wavelength is 50 meters. The receiving antenna is 10 kilometers from the transmitter and is oriented for maximum reception. It is high above the ground surface. Find the signal strength at the receiver in microvolts per meter, and the open-circuit voltage of the antenna in microvolts. Current at the base of the transmitting antenna is 1 ampere.

CHAPTER XIII

Wave Guides

Guided Waves. When an electric wave is sent out through empty space, it naturally becomes a spherical wave, for it travels with equal speed in all directions. There are no obstacles, and it can spread freely from its source. But the presence of material substance will affect its propagation.

The ability of a conducting surface to act as the boundary of an electric wave is used in various kinds of wave guides for directing the propagation of electric waves, as the rigid wall of a speaking tube is used to guide sound by preventing the sound wave from spreading into space. A power line or a telephone line is a wave guide; the surface of the copper wire provides a boundary on which the electric field of a wave can terminate and the wave propagates as a plane wave, following the conductor from end to end. Because of the conducting wires, the transmitted wave does not spread as a spherical wave but is able to travel as a plane wave, with its energy directed along the route of the transmission line.

To show the need for a boundary surface, consider the plane wave of Fig. 48, page 119. This wave was described in Chapter IX as extending without limit in a plane normal to its direction of travel. An unbounded wave like this is satisfactory from the mathematical point of view, but its energy would be infinite.

It is possible to put thin sheets of perfectly conducting material into the region through which a wave travels without affecting the wave, provided they are everywhere normal to the electric field. If conducting sheets were parallel or oblique to the electric field, current would flow in them that would distort the field, but conducting planes that are normal to the field, as in Fig. 68, will not disturb it.

When conducting surfaces are placed in the field, the electric lines of force will terminate on charge on the surfaces, and a plane wave can travel between such surfaces and be limited to the space between them. **It is not necessary** that the wave extend beyond the surfaces.

Consider the lower surface of Fig. 68. If a wave exists above it, but not below, as indicated, that surface must terminate the electric field and also the magnetic field. Electric field can terminate only on electric charge (for elsewhere its divergence is zero), so there must be an appropriate distribution of charge on the surface. Magnetic field can cease suddenly only at a surface carrying current (for elsewhere its curl is zero), so there must be current in the surface of Fig. 68. These two requirements are not independent, for, when the charge on the surface changes, as it must do to keep pace with the traveling wave, its motion constitutes current. This does not mean that charge flows as fast as the electric wave travels—air molecules transmitting a sound wave do not flow with the speed of sound—but current flows from regions of decreasing charge density to regions of increasing charge density.

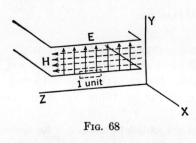

Fig. 68

If the plane wave is to move between conducting surfaces in a normal and undistorted manner, the currents flowing in the surfaces must accomplish these two requirements: they must at the same time provide necessary curl to act as a boundary for the magnetic field and provide the proper distribution of charge to terminate the electric field at all times. Let us see whether this is possible.

First, there is a definite relation between the electric field and the magnetic field of a wave, as defined by Maxwell's equations:

$$\nabla \times \mathbf{H} = \frac{\partial \mathbf{D}}{\partial t} \qquad [416]$$

Applying this equation to the plane wave of Fig. 68:

$$-\frac{\partial H_z}{\partial x} = \frac{\partial D_y}{\partial t} \qquad [417]$$

Second, using σ to represent charge per unit area on the lower conducting surface of Fig. 68:

$$\sigma = D_y \qquad [418]$$

As D changes with time, the charge density will change. Consider some point on the conducting surface at which the charge density is increasing; charge is supplied to this point by current that flows parallel to the X axis. If the current conveying this charge is flowing in the

GUIDED WAVES

positive X direction, is a positive current, and provides positive charge, it will be a little smaller after it has passed the point in question and deposited thereon some of its positive charge. Quantitatively, the decrease of current with respect to distance x is equal to the increase of charge with respect to time. If I_x is current in the X direction in a strip of the conducting surface of unit breadth,

$$-\frac{\partial I_x}{\partial x} = \frac{\partial \sigma}{\partial t} \qquad [419]$$

Combining this [1] with equation 418, the necessary relation between changing electric field and flow of current is

$$\frac{\partial D_y}{\partial t} = -\frac{\partial I_x}{\partial x} \qquad [420]$$

This tells the amount of current necessary to terminate the electric field.

Third, the current in the conducting surface of Fig. 68 must be related to the magnetic field strength parallel to the surface. As in equation 320

$$H_z = I_x \qquad [421]$$

By differentiation:

$$\frac{\partial H_z}{\partial x} = \frac{\partial I_x}{\partial x} \qquad [422]$$

Comparing equation 422 with 420, it appears that the same current will be satisfactory for terminating the electric field and for bounding the magnetic field if the electric and magnetic fields are related by

$$\frac{\partial H_z}{\partial x} = -\frac{\partial D_y}{\partial t} \qquad [423]$$

[1] Equation 419 is a special case of the "equation of continuity" which may be derived as follows. From Maxwell's equations,

$$\nabla \times \mathbf{H} = \mathbf{\iota} + \frac{\partial \mathbf{D}}{\partial t} \qquad [240]$$

Find the divergence of each side of this equation: on the left, the divergence of the curl is identically zero; on the right, there results a term for the divergence of $\mathbf{D}$. For the divergence of $\mathbf{D}$ substitute $\nabla \cdot \mathbf{D} = \rho$. Then

$$\nabla \cdot \mathbf{\iota} + \frac{\partial \rho}{\partial t} = 0$$

from which

$$\nabla \cdot \mathbf{\iota} = -\frac{\partial \rho}{\partial t}$$

196   WAVE GUIDES

But the electric and magnetic fields are always related in this manner, for this is identical with equation 417 derived from Maxwell's equation. The conclusion is that a wave *can* be limited to the space between two perfectly conducting parallel plane surfaces, since the current that flows in those surfaces will provide a proper boundary for both the electric and magnetic fields of the wave.

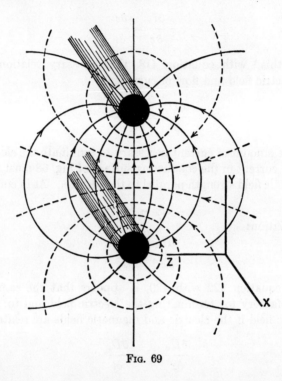

Fig. 69

To avoid the necessity of repeating this proof for every conducting surface that acts as a wave guide, it can be shown to be true in general. The proof is a generalization of the one given here for a plane wave, and the result will be accepted in our further discussion of wave guides.

Finite Waves. Although the wave of Fig. 68 is bounded in one dimension, it is still infinite in the other, and so it is not physically attainable. But the above discussion points the way to two types of wave guides that are actually useful.

Consider the lower conducting surface of Fig. 68 bent down and rolled into a cylinder, and the upper surface bent up and rolled into a cylinder. The result will be as shown in Fig. 69; there will be parallel conducting cylinders serving as a wave guide for a plane wave. The

FINITE WAVES

lines of electric flux will be bent and stretched into arcs of circles, and the magnetic flux lines will curve about the cylindrical conductors in an orthogonal set of circles.

To show that these conductors, which constitute a parallel-wire transmission line, will act as a satisfactory wave guide, it is necessary only to show that Maxwell's equations are satisfied in space between and around the conductors. This proof, although rather involved, is not difficult. For the simpler fields of a concentric transmission line, the proof gives no trouble. See Problem 1, page 226.

A wave guided by parallel wires is infinite in extent, but its strength diminishes in all directions, and its energy is finite. It is therefore a physically possible wave. The physical wave on such a line, however, differs from the mathematical wave in one way: it must be generated at one place and terminated at another, and near the ends of the transmission line the wave is not a plane wave. It goes through some kind of a transition, from quasi-spherical wave to quasi-plane wave. A similar disturbance takes place if there is a change of direction of the line, or a change of diameter or spacing of the conductors. When the wave is not strictly a plane wave, some energy is radiated away from the line and goes out into space generally, instead of following the conductors. This lost radiation is negligible at power frequency; at radio frequency it may be of great importance.

Most developments of transmission-line equations are based on the assumption that the capacitance and inductance of each short length of line may be considered independently of the rest of the line.[2] The assumption requires justification, for it is not apparent that there will be no electric or magnetic inductive effects between successive sections of the line. The justification is obtained in the proof outlined above: because the assumed electric and magnetic fields satisfy Maxwell's equations, they are correct. But this justification is obtained only with perfect conductors. If there is resistance, the transmission-line equations obtained by that method are good and useful approximations but not mathematically exact.

When the cylindrical conductors do not have perfect conductivity there is some penetration of current into the conductors; indeed, at low frequency, the current will penetrate the entire conductor. The speed of propagation is slightly less than the speed of light. The wave is not strictly a plane wave but is bent back near the conductor. Indeed, transmission-line equations must be considered as merely good approximations except when applied to lines of zero resistance. (This

[2] This development is given (with an explanatory note) in *Transient Electric Currents*, H. H. Skilling, Chapter IX.

is true even when line resistance and leakage are included in the equations.)

Hollow Wave Guides. A suggestion of another type of wave guide is obtained from consideration of the parallel planes of Fig. 68. A wave between two planes is restricted vertically but not horizontally. But let us use two other conducting planes as side walls of a rectangular wave guide as in Fig. 70.

A difficulty is immediately encountered. There cannot be any tangential electric field at the surface of these planes because they are conducting. Therefore it is necessary to consider a wave in which the

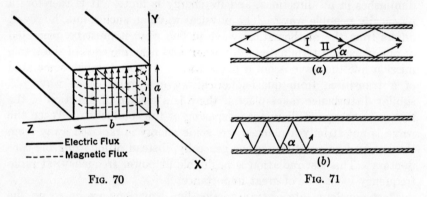

Fig. 70   Fig. 71

electric intensity, although everywhere vertical, diminishes to zero at the side surfaces of the guide. This is indicated in the diagram by decreased density of electric flux.

Such an electric field, because of its varying strength, has a component of curl along the X axis, and therefore requires an X component as well as a Z component of magnetic field. The resultant magnetic field is transverse at the middle of the wave guide, but it bends and becomes axial at the sides, as indicated by the dash lines in Fig. 70. The magnetic lines of flux form closed loops and the magnetic field is therefore without divergence, as indeed it must be.

The arrangement of electric and magnetic fields indicated in Fig. 70 cannot be produced by any single plane wave of the type that has been discussed in this and previous chapters, but it is rather surprising to find that it can be produced by two plane waves traveling within the wave guide at the same time. Neither of these waves goes axially along the guide, but each follows a zigzag path with multiple reflections from the walls of the guide as shown in Fig. 71a.

Only a sinusoidal wave will pass through a wave guide without distortion; let us therefore assume that waves I and II in Fig. 71a are sinusoidal waves. The angle at which they travel is dependent upon the wavelength and upon the size of the guide.

The essential point is this: two sinusoidal plane waves, of the same amplitude and frequency, when superimposed at an angle, will add to zero along certain surfaces.

Consider Fig. 72. Two waves are shown, one shaded black, the other white. They are traveling in somewhat different directions as shown by the arrows, the white wave almost directly away from the reader and the black wave toward the right. The waves are cut off at the near edge by a vertical plane through the dash line, the direction of which is midway between the directions of travel of the two waves.

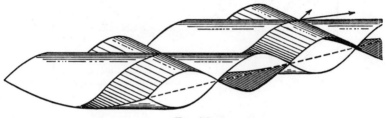

Fig. 72

The intersection of the waves with this vertical plane gives the pair of sine curves that appear at the edge of the waves in the figure.

The waves in this figure represent the electric intensity in two electromagnetic waves. Let the height of the pictured surfaces above or below the neutral plane be a representation of electric intensity. When the electric intensity of one wave is equal but opposite to the electric intensity of the other wave, the resultant electric intensity is zero. This is true all along the dash line, for, where the black wave is above the dash line, the white one is below by an equal amount and vice versa.

This is also the situation along another plane, parallel to the dash line, that cuts off the waves on the farther side. Along both of these planes, the sum of the two waves is always exactly zero. At intermediate points, the sum of the two waves is not zero, and the total resultant wave has the shape shown in Fig. 73, with a maximum value (where the black and white crests of Fig. 72 coincide) of twice the crest of either component wave alone.

Because the resultant electric field strength at the indicated boundary surfaces will always be zero, the pair of electromagnetic waves represented by the diagram may be contained within a rectangular wave guide of proper dimensions. They satisfy the requirement that there be no tangential component of electric field at a conducting surface:

along the top and bottom of the guide the electric field is normal to the surface, and along the two side walls it is zero. Both waves follow zigzag paths within the guide, as in Fig. 71, and are reflected back and forth from side to side. In fact, there is essentially but one wave, for each wave is the reflection of the other.

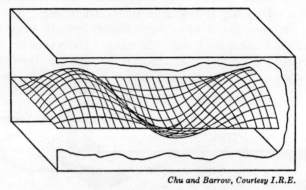

Chu and Barrow, Courtesy I.R.E.

Fig. 73

The total result is that a series of electromagnetic pulses go down the wave guide as in Fig. 73. This diagram shows the nature of the electric component of the wave, but it does not indicate the magnetic field. Figure 74 shows the distribution of both electric and magnetic components in this type of guided wave (known as the $TE_{0,1}$ wave).

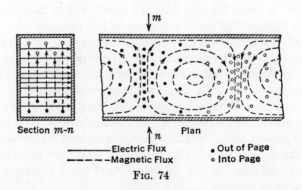

Fig. 74

Group Velocity. Waves in hollow guides have a number of peculiar properties and limitations. One is that the transmission of a radio signal along the tube will be at a speed somewhat less than the speed of light. This is easily explained in terms of the zigzag path followed by the two components, as shown in Fig. 71. Each plane-wave component travels at the speed of light along its zigzag path, but the rate at which a signal travels *along the guide* will be somewhat less because the zigzag path is longer than the axis of the guide. If a telegraphic dot or dash, or a telephonic modulation envelope could be observed, it would be seen to be going along the guide a little slower than the speed of light. Its velocity is known as the **group velocity,** v_g.

The group velocity of a wave in a guide is dependent upon its frequency. Consider crossed waves, as in Fig. 75, proceeding along a

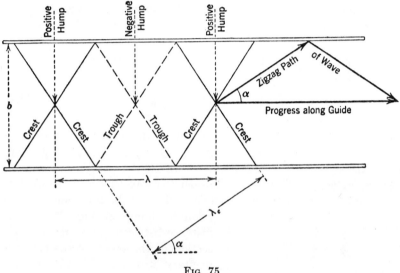

Fig. 75

rectangular wave guide. The direction of travel is normal to the crests, as shown. To make the electric field zero along both sides of the guide, as in Fig. 72, the elementary waves must cross at such an angle that they overlap one wavelength in the width of the guide. Thus wavelength λ_e and guide width b determine the angle of the elementary waves. The greater the wavelength of the component waves, the greater the angle α. For a low-frequency wave, the zigzag path will be squeezed up as in Fig. 71b, and group velocity will be relatively low. Finally, if the half wavelength of the wave becomes as great as the width of the wave guide, the two component waves will be reflected back and forth across the guide, giving a standing wave, but there will be no forward progress at all. This is the **cut-off frequency** for the wave guide, and a wave of lower frequency cannot be transmitted because no wave of lower frequency can have zero electric field at both side walls of the guide.

The longest wavelength that can be transmitted along a rectangular guide with dimensions a and b as in Fig. 70 is therefore

$$\lambda_0 = 2b \qquad [424]$$

and the cut-off frequency is

$$f_0 = \frac{v_e}{2b} = \frac{1}{2b\sqrt{\mu\epsilon}} \qquad [425]$$

Note that, in equation 424, λ_0 is the cut-off wavelength of the *elementary* wave, or the wavelength of an *unbounded* wave at the cut-off frequency, and that correspondingly the velocity of propagation of an unbounded wave, $v_e = 1/\sqrt{\mu\epsilon}$, is used in equation 425.

The group velocity of the $TE_{0,1}$ wave is, from Fig. 75:

$$v_g = v_e \sqrt{1 - \left(\frac{\lambda_e}{2b}\right)^2} \qquad [426]$$

It is dependent upon λ_e, the wavelength of the elementary wave. It will be seen to approach zero at cut-off, as λ_e approaches $2b$, and to approach the speed of light for very short waves.

Phase Velocity. Next let us consider the wavelength of the electromagnetic field within the wave guide. This will be the distance from hump to hump of Fig. 73, or the distance between the points of intersection of the crests of the two elementary waves in Fig. 72 or 75. Wave crests are indicated in Fig. 75; in this diagram, the distance λ_e is the length of the elementary waves, and λ is the apparent wavelength or the distance between humps of the total wave. The apparent wavelength is greater than the elementary wavelength.

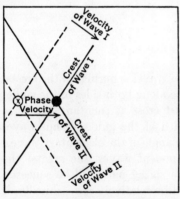

Fig. 76

Corresponding to this apparent wavelength is an apparent velocity greater than the speed of light. Consider the crossed crests of the two elementary waves as shown in Fig. 76. The solid lines show the present positions of the crests; the dash lines show their positions one unit of time earlier. The velocity is given by the distance through which the waves advance in this interval. At the point of intersection of the waves is the crest of the resultant or apparent wave. This point of intersection advances more rapidly than do the individual waves. The result is that if the wave pattern of Fig. 73 were visible it would appear to travel along the wave guide at a velocity greater than that of light. This apparent velocity is called the **phase velocity,** for it is the rate at which a given phase of the resultant wave travels.

It seems paradoxical that the waves should appear to travel in the guide at a speed greater than that of light, while a signal conveyed by those waves would go at the group velocity which is less than the speed of light, yet that is what happens. Consider Fig. 77, which shows a "carrier" wave contained within a modulation envelope.[3] If it were

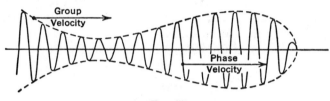

Fig. 77

visible in a wave guide, the carrier wave would appear to be advancing at *phase velocity,* while the modulation pattern would progress at the slower *group velocity.*[4] The carrier wave would therefore seem to be slipping forward through the modulation envelope, and each individual crest of the carrier wave would change amplitude as it passed through the irregularities of the envelope until it disappeared entirely upon reaching the most advanced point of the envelope.

The manner in which the apparent wave can pass out of existence is illustrated by Fig. 76. The wave crests in that figure come to an end toward the right. That may be considered the farthest point to which the signal has advanced. The apparent wave is the resultant of the two elementary waves and appears as a hump at the point of intersection of the elementary waves; the hump advances rapidly from left to right until the elementary wave crests have passed beyond each other and no longer intersect, and then it simply disappears. Consider two waves traveling at an angle to each other on the surface of a body of water, and the disappearance of the double crest at their point of intersection is easily visualized.

[3] Consider also, as a thoroughly non-mathematical example, a hurrying caterpillar. Little waves (transverse or compressional, depending on the species) ripple along its back from tail to head. These waves progress at phase velocity, while the caterpillar himself travels at group velocity.

[4] See Chapter XIV, pages 228 to 231.

An ocean wave, or a large wave on a lake, often approaches a retaining wall along the shore at a slight angle. Instead of advancing squarely upon the shore so that its full length breaks upon the wall at the same instant, the wave reaches the wall at one end a little sooner than it does at the other. When one watches such a wave, there is a surge and splash at the point where the wave is breaking against the wall. This surge appears at one end of the wall with the first arrival of the wave, and then, as the wave breaks progressively from one end of the wall to the other, the surge (which marks the wave crest) appears to travel with great speed along the wall. Its speed may appear many times that of the actual rate of advance of the wave, and the less the angle between wave and wall the more rapidly it will seem to go. This apparent speed along the wall is a phase velocity, exactly analogous to the phase velocity in a wave guide.

The phase velocity of the $TE_{0,1}$ wave is, from Fig. 76:

$$v_\phi = \frac{v_e}{\sqrt{1 - \left(\frac{\lambda_e}{2b}\right)^2}} \qquad [427]$$

and it will be seen that the phase velocity is greater than the velocity of light in the same proportion as the group velocity is less. For very short waves the phase velocity, like the group velocity, approaches the speed of light, but at frequencies near cut-off the phase velocity approaches infinity.

Note particularly that a signal cannot travel at phase velocity but only at group velocity.

Derivation. It is obvious that, whether one looks upon the phenomena within a wave guide as a single wave or as a pair of zigzag waves, the electric and magnetic fields must be consistent with Maxwell's equations.

Since the electric field of the wave of Fig. 70 is parallel to the Y axis, the X and Z components may at once be set equal to zero as in equations 429 below.

The wave is assumed to be sinusoidal with respect to time, and, since it is traveling along the X axis, it will be described by a sinusoidal function of $(x - v_\phi t)$ where v_ϕ is its phase velocity. If its frequency is f, and $\omega = 2\pi f$, the electric field may be written as the following function of time:

$$\sin \frac{\omega}{v_\phi}(x - v_\phi t) = \sin(\beta x - \omega t) \qquad [428]$$

DERIVATION

The electric field must be zero at $z = 0$ and at $z = b$. One way to accomplish this is to have a sinusoidal variation of electric field strength along the Z axis, starting at zero, rising to a maximum value at $b/2$, and falling again to zero at the side of the wave guide where $z = b$. With this sinusoidal space distribution, if the maximum strength of the field is A its strength at any value of z is $A \sin \dfrac{z}{b} \pi$.

Combining the variation-with-time and the variation-from-side-to-side-of-the-guide in a single equation

$$E_y = A \sin\left(\frac{z}{b}\pi\right) \sin(\beta x - \omega t)$$

and [429]

$$E_x = 0$$

$$E_z = 0$$

The wave of equation 429 satisfies the boundary conditions imposed by the rectangular guide, for the equation was written with that in mind, but it is a physically possible electromagnetic wave only if it is a solution of the wave equation

$$\nabla^2 \mathbf{E} = \mu\epsilon \frac{\partial^2 \mathbf{E}}{\partial t^2} \qquad [249]$$

and if

$$\nabla \cdot \mathbf{E} = 0 \qquad [244]$$

The latter condition is obviously satisfied. When the wave equation is expanded in rectangular coordinates with $E_x = 0$ and $E_z = 0$, there results simply (see equation 250):

$$\frac{\partial^2 E_y}{\partial x^2} + \frac{\partial^2 E_y}{\partial y^2} + \frac{\partial^2 E_y}{\partial z^2} = \mu\epsilon \frac{\partial^2 E_y}{\partial t^2} \qquad [430]$$

Differentiating E_y from equation 429, substituting into equation 430, and simplifying, one obtains

$$\frac{\pi^2}{b^2} + \beta^2 = \mu\epsilon\omega^2 \qquad [431]$$

The conclusion is that the wave proposed in equation 429 is a satisfactory solution of the wave equation *if* equation 431 can be satisfied.

Noting that $\beta = \omega/v_\phi$, equation 431 can be satisfied by a wave with phase velocity given by

$$v_\phi{}^2 = \frac{1}{\mu\epsilon - \left(\dfrac{\pi}{\omega b}\right)^2} \qquad [432]$$

This is interpreted to mean that the proposed wave can exist only if it travels with a phase velocity dependent upon the frequency of the wave and the size of the guide. If the frequency is so low that $v_\phi{}^2$ is negative, no real solution for phase velocity is possible, the wave equation cannot then be satisfied, and the wave cannot exist. This gives the cut-off frequency, which will be found to agree with equation 425. Equation 432, giving the phase velocity, can be transformed to correspond with equation 427.

It is now known that the electric wave can travel in the wave guide. It can have any amplitude, A, in equation 429, but its frequency must exceed a certain minimum, and its phase velocity is determined by the frequency and the size of the guide.

To determine the magnetic field of the wave, we use Maxwell's equation:

$$\nabla \times \mathbf{E} = -\mu \frac{\partial \mathbf{H}}{\partial t} \qquad [242]$$

Finding partial derivatives of E_y from equation 429 and substituting into equation 242,

$$\frac{\partial E_y}{\partial x} = -\mu \frac{\partial H_z}{\partial t} = \beta A \sin\left(\frac{z}{b}\pi\right) \cos(\beta x - \omega t)$$

$$-\frac{\partial E_y}{\partial z} = -\mu \frac{\partial H_x}{\partial t} = -\frac{\pi}{b} A \cos\left(\frac{z}{b}\pi\right) \sin(\beta x - \omega t) \qquad [433]$$

The components of magnetic field are found by integrating the second and third members of these equations with respect to time. This gives

$$H_z = \frac{1}{\mu v_\phi} A \sin\left(\frac{z}{b}\pi\right) \sin(\beta x - \omega t)$$

$$H_x = \frac{\pi}{\mu\omega b} A \cos\left(\frac{z}{b}\pi\right) \cos(\beta x - \omega t) \qquad [434]$$

$$H_y = 0$$

All components of the electric and magnetic fields of the guided wave are now known, and consideration of equations 429 and 434 shows that Fig. 74 is a correct representation of the wave. The line m–n in that figure is through a section of the wave at which $(\beta x - \omega t)$ is very slightly less than $\pi/2$.

General Solution for Rectangular Guide. The previous discussion has been concerned with one mode of wave propagation only. This, called the "dominant mode" for a rectangular guide, is the simplest and perhaps the most useful, but a rectangular guide is able to carry a variety of modes. A general solution of Maxwell's equations should yield all possible modes.

Let us consider a rectangular guide as in Fig. 78. Guided waves may in general have field components in all three coordinate directions. It is always possible to consider the most general wave as the sum of two waves, one of which has only transverse and no axial magnetic field, whereas the other has only transverse and no axial electric field. The first kind is called a **Transverse Magnetic** or TM wave, the latter a **Transverse Electric** or TE wave. Some waves, as on a transmission line, having no axial components of either field, are called TEM waves.

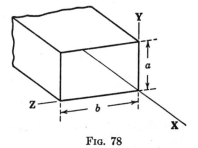

Fig. 78

To be fully general, assume the electric field in the guide has three components, E_x, E_y, and E_z, each of which is an undefined function of x, y, z, and t. Although these components are as yet undefined, they must satisfy three conditions. They must satisfy the wave equation:

$$\nabla^2 \mathbf{E} = \mu\epsilon \frac{\partial^2 \mathbf{E}}{\partial t^2} \qquad [435]$$

They must satisfy the divergence equation:

$$\nabla \cdot \mathbf{E} = 0 \qquad [436]$$

They must satisfy the boundary conditions: that is, if the walls of the guide are assumed perfectly conducting, with dimensions as in Fig. 78, it is necessary that

At $z = 0$ and at $z = b$ $\qquad E_x = E_y = 0 \qquad [437]$

At $y = 0$ and at $y = a$ $\qquad E_x = E_z = 0 \qquad [438]$

The following solution is to find functions that satisfy these requirements.

First, let us assume the electric field may be a sinusoidal wave traveling along the guide and determine whether such an assumption is consistent with the necessary conditions. We write the components of the traveling wave as:

$$E_x = E_{mx}\, e^{j\omega t - \Gamma x}$$
$$E_y = E_{my}\, e^{j\omega t - \Gamma x} \qquad [439]$$
$$E_z = E_{mz}\, e^{j\omega t - \Gamma x}$$

where E_{mx}, E_{my}, and E_{mz} are functions of y and z but not of x or t.

Giving attention to E_x, the wave equation requires that

$$\frac{\partial^2 E_x}{\partial x^2} + \frac{\partial^2 E_x}{\partial y^2} + \frac{\partial^2 E_x}{\partial z^2} = \mu\epsilon \frac{\partial^2 E_x}{\partial t^2} \qquad [440]$$

Introducing E_x from equation 439 and dividing out the exponential function,

$$\frac{\partial^2 E_{mx}}{\partial y^2} + \frac{\partial^2 E_{mx}}{\partial z^2} = -(\omega^2 \mu\epsilon + \Gamma^2) E_{mx} \qquad [441]$$

A solution of this partial differential equation is found by separating the variables into two ordinary differential equations. Write E_{mx} (which is a function of y and z) as the product of two functions: Y, a function of y only, and Z, a function of z only. Then $E_{mx} = YZ$, and substitution into equation 441 gives

$$Z \frac{d^2 Y}{dy^2} + Y \frac{d^2 Z}{dz^2} = -(\omega^2 \mu\epsilon + \Gamma^2) YZ \qquad [442]$$

whence

$$\frac{1}{Y} \frac{d^2 Y}{dy^2} + \frac{1}{Z} \frac{d^2 Z}{dz^2} = -(\omega^2 \mu\epsilon + \Gamma^2) \qquad [443]$$

We may arbitrarily let

$$\frac{1}{Y} \frac{d^2 Y}{dy^2} = -M^2 \quad \text{and} \quad \frac{1}{Z} \frac{d^2 Z}{dz^2} = -N^2 \qquad [444]$$

from which

$$M^2 + N^2 = \omega^2 \mu\epsilon + \Gamma^2 \qquad [445]$$

Since the right-hand member of this equation is constant (not a function of y or z), M and N must also be constants; this makes equations

GENERAL SOLUTION FOR RECTANGULAR GUIDE 209

444 ordinary differential equations with the following solutions [5] (which, if unfamiliar, may be checked by substitution):

$$Y = A_1 \cos My + B_1 \sin My$$
$$Z = C_1 \cos Nz + D_1 \sin Nz \qquad [446]$$

The product, then, is

$$E_{mx} = (A_1 \cos My + B_1 \sin My)(C_1 \cos Nz + D_1 \sin Nz) \qquad [447]$$

where A_1, B_1, C_1, and D_1 are any constants.

Similar solutions of the wave equation result for E_{my} and E_{mz}, with other arbitrary constants.

Using equations 437 and 438 to introduce boundary conditions, the constants of equation 447 will be somewhat limited. $E_x = 0$ at $z = 0$ requires that $C_1 = 0$, and at $y = 0$ requires that $A_1 = 0$. Then E_x will be zero at $z = b$ if Nb is some integer multiple of π; that is, if

$$N = \frac{n\pi}{b} \quad \text{where } n = 0, 1, 2, 3, \cdots \qquad [448]$$

Likewise, to make $E_x = 0$ at $y = a$,

$$M = \frac{m\pi}{a} \quad \text{where } m = 0, 1, 2, 3, \cdots \qquad [449]$$

Hence, when boundary conditions are satisfied, we have

$$E_{mx} = B_1 D_1 \sin \frac{m\pi}{a} y \sin \frac{n\pi}{b} z \qquad [450]$$

B_1 and D_1 are both arbitrary constants and can be combined as K_1. Then

$$E_x = K_1 \sin \frac{m\pi}{a} y \sin \frac{n\pi}{b} z \, e^{j\omega t - \Gamma x} \qquad [451]$$

Introducing boundary conditions into the similar solutions for the other components gives

$$E_y = (A_2 \cos M_2 y + B_2 \sin M_2 y)\left(D_2 \sin \frac{n\pi}{b} z\right) e^{j\omega t - \Gamma x} \qquad [452]$$

$$E_z = \left(B_3 \sin \frac{m\pi}{a} y\right)(C_3 \cos N_3 z + D_3 \sin N_3 z) e^{j\omega t - \Gamma x} \qquad [453]$$

[5] Solutions may be expressed in hyperbolic or exponential functions if preferred. The trigonometric form is here more convenient. Note the similarity to transmission line equations.

WAVE GUIDES

The divergence equation requires that

$$\frac{\partial E_x}{\partial x} + \frac{\partial E_y}{\partial y} + \frac{\partial E_z}{\partial z} = 0$$

There is no chance of satisfying this requirement unless B_2 and D_3 equal zero, and $M_2 = m\pi/a$, and $N_3 = n\pi/b$. With these stipulations, however, the derivatives are

$$\frac{\partial E_x}{\partial x} = -\Gamma K_1 \sin\frac{m\pi}{a} y \sin\frac{n\pi}{b} z \, e^{j\omega t - \Gamma x}$$

$$\frac{\partial E_y}{\partial y} = -\frac{m\pi}{a} K_2 \sin\frac{m\pi}{a} y \sin\frac{n\pi}{b} z \, e^{j\omega t - \Gamma x} \qquad [454]$$

$$\frac{\partial E_z}{\partial z} = -\frac{n\pi}{b} K_3 \sin\frac{m\pi}{a} y \sin\frac{n\pi}{b} z \, e^{j\omega t - \Gamma x}$$

and the divergence equation is satisfied if

$$\Gamma K_1 + \frac{m\pi}{a} K_2 + \frac{n\pi}{b} K_3 = 0 \qquad [455]$$

It is now possible to write the electric field components as

$$E_x = K_1 \sin\frac{m\pi}{a} y \sin\frac{n\pi}{b} z \, e^{j\omega t - \Gamma x}$$

$$E_y = K_2 \cos\frac{m\pi}{a} y \sin\frac{n\pi}{b} z \, e^{j\omega t - \Gamma x} \qquad [456]$$

$$E_z = K_3 \sin\frac{m\pi}{a} y \cos\frac{n\pi}{b} z \, e^{j\omega t - \Gamma x}$$

with the K's limited by equation 455, and with Γ from equation 445:

$$\Gamma^2 = \left(\frac{m\pi}{a}\right)^2 + \left(\frac{n\pi}{b}\right)^2 - \omega^2 \mu \epsilon \qquad [457]$$

Magnetic field components in the guide are found from Maxwell's equation

$$\frac{\partial \mathbf{H}}{\partial t} = -\frac{1}{\mu} \nabla \times \mathbf{E} \qquad [458]$$

TRANSVERSE ELECTRIC WAVES

This familiar operation is applied to equations 456 and gives

$$H_x = \frac{\pi}{j\omega\mu}\left(\frac{n}{b}K_2 - \frac{m}{a}K_3\right)\cos\frac{m\pi}{a}y\cos\frac{n\pi}{b}z\, e^{j\omega t - \Gamma x}$$

$$H_y = \frac{-\pi}{j\omega\mu}\left(\frac{n}{b}K_1 + \frac{\Gamma}{\pi}K_3\right)\sin\frac{m\pi}{a}y\cos\frac{n\pi}{b}z\, e^{j\omega t - \Gamma x} \quad [459]$$

$$H_z = \frac{\pi}{j\omega\mu}\left(\frac{m}{a}K_1 + \frac{\Gamma}{\pi}K_2\right)\cos\frac{m\pi}{a}y\sin\frac{n\pi}{b}z\, e^{j\omega t - \Gamma x}$$

Transverse Electric Waves. Equations 456 and 459 cannot be used until one more fact about the wave to be propagated is specified, for they contain too many unknown coefficients. The additional fact might be the relative magnitudes of E_x and H_x; as a highly important special case let us specify that $E_x = 0$. The wave is then a transverse electric or TE wave. This requires that $K_1 = 0$. Making K_1 zero in equations 455, 456, and 459 gives the equations of Table VI. Two new symbols require explanation: A_0, the only remaining arbitrary constant, determines the amplitude of each component and hence of the whole wave. Physically, A_0 depends on the size of the wave sent into the guide; mathematically, $A_0 = K_2 b/n$. G is an abbreviation for a combination of constants that appears frequently; it relates the characteristic factors of propagation in the guide to the corresponding factors in unbounded space, and

$$G^2 = 1 - \frac{1}{4f^2\mu\epsilon}\left(\frac{m^2}{a^2} + \frac{n^2}{b^2}\right) \quad [460]$$

As in previous chapters, η is the intrinsic impedance of the dielectric within the wave guide (usually air) and, if there is no dielectric loss, $\eta = \sqrt{\mu/\epsilon}$.

TE$_{m,n}$ TABLE VI

Components of Transverse Electric Waves in Rectangular Guides

$$E_x = 0$$

$$E_y = A_0 \frac{n}{b}\cos\frac{m\pi}{a}y\sin\frac{n\pi}{b}z\, e^{j\omega t - \Gamma x}$$

$$E_z = -A_0 \frac{m}{a}\sin\frac{m\pi}{a}y\cos\frac{n\pi}{b}z\, e^{j\omega t - \Gamma x}$$

$$H_x = \frac{A_0}{j2f\mu}\left(\frac{m^2}{a^2} + \frac{n^2}{b^2}\right)\cos\frac{m\pi}{a}y\cos\frac{n\pi}{b}z\, e^{j\omega t - \Gamma x}$$

$$H_y = -\frac{G}{\eta}E_z$$

$$H_z = \frac{G}{\eta}E_y$$

(See also Table VIII)

TM$_{m,n}$

TABLE VII

Components of Transverse Magnetic Waves in Rectangular Guides

$$H_x = 0$$

$$H_y = B_0 \frac{n}{b} \sin \frac{m\pi}{a} y \cos \frac{n\pi}{b} z \, e^{j\omega t - \Gamma x}$$

$$H_z = -B_0 \frac{m}{a} \cos \frac{m\pi}{a} y \sin \frac{n\pi}{b} z \, e^{j\omega t - \Gamma x}$$

$$E_x = \frac{B_0}{j2f\epsilon} \left(\frac{m^2}{a^2} + \frac{n^2}{b^2}\right) \sin \frac{m\pi}{a} y \sin \frac{n\pi}{b} z \, e^{j\omega t - \Gamma x}$$

$$E_y = \eta G H_z$$

$$E_z = -\eta G H_y$$

(See also Table VIII)

TABLE VIII

Auxiliary Formulas for Both TE and TM Waves in Rectangular Guides

$$G = \sqrt{1 - \frac{1}{4f^2 \mu \epsilon}\left(\frac{m^2}{a^2} + \frac{n^2}{b^2}\right)} = \sqrt{1 - \frac{9 \times 10^{16}}{4f^2}\left(\frac{m^2}{a^2} + \frac{n^2}{b^2}\right)} \text{ in air (see note)}$$

$$\eta = \sqrt{\frac{\mu}{\epsilon}} \qquad = 377 \text{ ohms in air}$$

$$\Gamma = j\omega\sqrt{\mu\epsilon}\,G \qquad = j\frac{2\pi f G}{3 \times 10^8} \text{ per meter in air}$$

$$\beta = \omega\sqrt{\mu\epsilon}\,G \qquad = \frac{2\pi f G}{3 \times 10^8} \text{ radians per meter in air}$$

$$v_g = \frac{G}{\sqrt{\mu\epsilon}} \qquad = 3G \times 10^8 \text{ meters per second in air}$$

$$v_\phi = \frac{1}{\sqrt{\mu\epsilon}\,G} \qquad = \frac{3 \times 10^8}{G} \text{ meters per second in air}$$

$$\lambda = \frac{1}{\sqrt{\mu\epsilon}fG} \qquad = \frac{3 \times 10^8}{fG} \text{ meters (wavelength in guide)}$$

Cut-off frequency: $G = 0$

Note: The right-hand column is for a hollow or air-dielectric guide.
In addition to the usual symbols, in Tables VI, VII, and VIII:

a is the dimension of the guide in the y direction, in meters.
m is the number of half-cycles of field pattern in the guide in the y direction, an integer. Thus a/m is the length of a half-cycle in the y direction.
b is the dimension of the guide in the z direction, in meters.
n is the number of half-cycles of field pattern in the guide in the z direction, an integer. Thus b/n is the length of a half-cycle in the z direction.
A_0 and B_0 are arbitrary constants, giving amplitude.
f is frequency in cycles per second.

Transverse Magnetic Waves. If, on the other hand, it is specified that $H_x = 0$, giving a TM wave, equation 459 shows that

$$\frac{n}{b} K_2 - \frac{m}{a} K_3 = 0 \qquad [461]$$

Using this with equation 455 in the general equations 456 and 459, there result the forms in Table VII.

Modes. The equations of Table VI (or VII) describe not a single wave but many. Both m and n may have any integer value. (For a

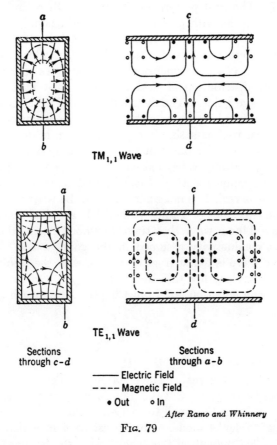

Fig. 79

TE wave, either m or n may be zero, but if both are zero the wave vanishes; for a TM wave, if either is zero the wave vanishes.)

Each pair of integers m,n corresponds to what is called a mode of propagation. It is customary to distinguish the various possible modes as $TE_{m,n}$ or $TM_{m,n}$ modes. Thus, if $m = 0$ and $n = 1$ in Table VI, the $TE_{0,1}$ mode results, and this will be recognized as the mode discussed in the early part of the chapter and shown in Fig. 74. (This differs from the $TE_{1,0}$ mode only in orientation.) The formulas of Tables VI and VIII for the $TE_{0,1}$ mode will be found to agree with equations 425 to 434. The $TE_{1,1}$ and $TM_{1,1}$ modes are illustrated in Fig. 79. All other rectangular modes result from repeating one of these three basic patterns a number of times within the cross section of a guide.

Any number of modes can exist in a wave guide at the same time, superimposed. The field pattern in the guide is therefore not limited to sinusoidal variation with y and z. The equations of Tables VI and VII could properly be written as summations, each having a doubly infinite number of terms as m and n take all integer values successively. Such a summation is a two-dimensional Fourier series and can be used to describe any distribution of field that may exist in a given cross section of the wave guide. The distribution in all other cross sections can then be computed from the equations. A non-sinusoidal pattern will *not* be propagated as a simple traveling wave, however; it will be distorted as it travels, for different modes are propagated at different phase velocities, and modes beyond cut-off will not propagate at all.

Cut-Off. The guide constant G can be real or imaginary (but not complex). If G is real, Γ is imaginary and can be written $\Gamma = j\beta$ with a real β. The equations of Tables VI and VII are then traveling-wave equations, describing a wave that travels without attenuation (a result of the perfect conductivity assumed in the guide walls). The phase velocity of the wave is found from $v_\phi = \omega/\beta$, with the result shown in Table VIII. Wavelength within the guide is v_ϕ/f. Group velocity can be found by using equation 467 from Chapter XIV, which gives

$$v_g = \frac{d\omega}{d\beta} = \frac{G}{\sqrt{\mu\epsilon}} \qquad [462]$$

and these results also are in Table VIII.

If G is imaginary, Γ is real. The equations of Tables VI and VII do not then describe traveling waves. Physically a real value of Γ indicates that the guide is too small to carry a wave of such low frequency or of such high order of mode. The lowest frequency that can travel in a guide is called the cut-off frequency and is the frequency at which $G = 0$, for any lower frequency would make G imaginary. Thus, the cut-off frequency is

$$f_0 = \sqrt{\frac{1}{4\mu\epsilon}\left(\frac{m^2}{a^2} + \frac{n^2}{b^2}\right)} \qquad [463]$$

Below cut-off, an electromagnetic field introduced into the guide will not produce a wave. The equations of Tables VI and VII show that instead there will be a field in the guide that diminishes exponentially with distance x along the guide, and that alternates in the same phase all along the guide. Such a field does not convey energy along the guide, and (assuming perfectly conducting walls) the average power entering the guide below cut-off is zero. The exponential decrease of field strength in such a guide is not attenuation in the ordinary sense, for it is not the result of loss of energy. It is rather like the "attenuation" in a lossless filter circuit and is sometimes called "reactive attenuation."

From the ambiguity of sign of equation 457, Γ and hence β may be either positive or negative. Positive β gives wave propagation in the positive x direction; negative β in the negative x direction. As a matter of fact, there may be propagation in either direction, or in both at once, depending on conditions at the ends of the wave guide. For instance, if energy is supplied at one end and there is a discontinuity that produces partial reflection at the other end, both incident and reflected waves will exist within the guide. They will combine to give standing waves along the guide. Such standing waves are of great practical importance, for they indicate some degree of mismatch at the discontinuity and a consequent restriction of flow of energy in the system.

Since G, the guide factor, is always less than one, phase velocity in a guide is always greater than the velocity of an unbounded wave, and group velocity in the guide is correspondingly less. This appears from Table VIII. For all hollow rectangular guides (and, indeed, for all hollow guides of any shape)

$$v_\phi v_g = \frac{1}{\mu\epsilon} = c^2 \qquad [464]$$

Cylindrical Wave Guides. The common shape of guide, other than rectangular, is of circular cross section. The solution for waves in such guides is very similar to that for waves in rectangular guides. However, one of the equations for a cylindrical guide, corresponding to equation 444, turns out to have a solution in the form of a Bessel function instead of a trigonometric function.

Using cylindrical coordinates with the Z axis coinciding with the axis of the wave guide, the r, θ, and z components of field are found. The general form of solution for all modes will not be given, but four of the more interesting modes are described in Table IX and illustrated in Fig. 80. In Table IX, a is radius of the guide, J_0 and J_1 are Bessel functions,[6] A is the constant that gives amplitude of the wave, λ is the

---

[6] Values may be obtained from mathematical handbooks, such as *Tables of Functions*, Jahnke and Emde, Dover Publications, 1943.

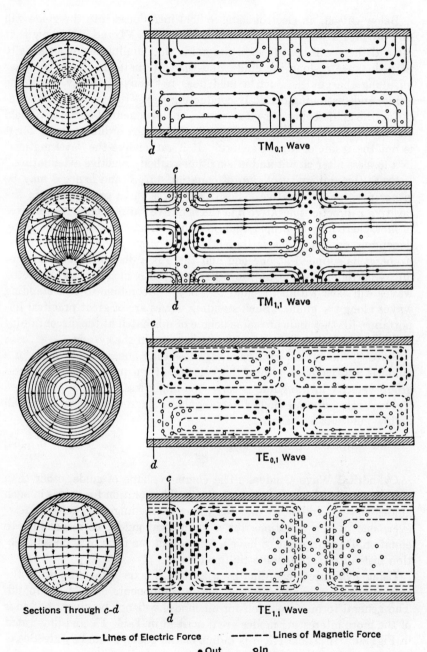

Sections Through c-d   TM₀,₁ Wave / TM₁,₁ Wave / TE₀,₁ Wave / TE₁,₁ Wave

——— Lines of Electric Force   - - - - Lines of Magnetic Force
● Out   ○ In

G. C. Southworth, Bell Labs, *courtesy Electrical Engineering*

FIG. 80

TABLE IX
Certain Modes in Cylindrical Guides

$TM_{0,1}$ Mode	$TM_{1,1}$ Mode
$E_z = AJ_0\left(2.405\dfrac{r}{a}\right)e^{j\omega t-\Gamma z}$	$E_z = AJ_1\left(3.83\dfrac{r}{a}\right)\cos\theta\, e^{j\omega t-\Gamma z}$
$E_r = \eta G H_\theta$	$E_r = \eta G H_\theta$
$E_\theta = 0$	$E_\theta = -\eta G H_r$
$H_z = 0$	$H_z = 0$
$H_r = 0$	$H_r = -j0.429\dfrac{a^2 f\epsilon}{r}AJ_1\left(3.83\dfrac{r}{a}\right)\cdot\sin\theta\, e^{j\omega t-\Gamma z}$
$H_\theta = j2.61\, af\epsilon AJ_1\left(2.405\dfrac{r}{a}\right)e^{j\omega t-\Gamma z}$	$H_\theta = -j1.64 af\epsilon A\left[J_0\left(3.83\dfrac{r}{a}\right) - \dfrac{a}{3.83r}J_1\left(3.83\dfrac{r}{a}\right)\right]\cos\theta\, e^{j\omega t-\Gamma z}$
$G = \sqrt{1-\dfrac{0.147}{f^2 a^2\mu\epsilon}}$	$G = \sqrt{1-\dfrac{0.371}{f^2 a^2\mu\epsilon}}$
$f_0 = \dfrac{0.383}{a\sqrt{\mu\epsilon}}$	$f_0 = \dfrac{0.609}{a\sqrt{\mu\epsilon}}$

$TE_{0,1}$ Mode	$TE_{1,1}$ Mode
$E_z = 0$	$E_z = 0$
$E_r = 0$	$E_r = j1.86\dfrac{a^2 f\mu}{r}AJ_1\left(1.84\dfrac{r}{a}\right)\cdot\sin\theta\, e^{j\omega t-\Gamma z}$
$E_\theta = -j1.64 af\mu AJ_1\left(3.83\dfrac{r}{a}\right)e^{j\omega t-\Gamma z}$	$E_\theta = j3.41 af\mu A\left[J_0\left(1.84\dfrac{r}{a}\right) - \dfrac{a}{1.84r}J_1\left(1.84\dfrac{r}{a}\right)\right]\cos\theta\, e^{j\omega t-\Gamma z}$
$H_z = AJ_0\left(3.83\dfrac{r}{a}\right)e^{j\omega t-\Gamma z}$	$H_z = AJ_1\left(1.84\dfrac{r}{a}\right)\cos\theta\, e^{j\omega t-\Gamma z}$
$H_r = -\dfrac{G}{\eta}E_\theta$	$H_r = -\dfrac{G}{\eta}E_\theta$
$H_\theta = 0$	$H_\theta = \dfrac{G}{\eta}E_r$
$G = \sqrt{1-\dfrac{0.371}{f^2 a^2\mu\epsilon}}$	$G = \sqrt{1-\dfrac{0.0858}{f^2 a^2\mu\epsilon}}$
$f_0 = \dfrac{0.609}{a\sqrt{\mu\epsilon}}$	$f_0 = \dfrac{0.293}{a\sqrt{\mu\epsilon}}$

For All Modes

$\Gamma = j\omega\sqrt{\mu\epsilon}\,G \qquad \lambda = \dfrac{1}{\sqrt{\mu\epsilon}fG} \qquad v_\phi = \dfrac{1}{\sqrt{\mu\epsilon}\,G} \qquad a =$ radius of guide

$\beta = \omega\sqrt{\mu\epsilon}\,G \qquad \eta = \sqrt{\mu/\epsilon} \qquad v_g = \dfrac{G}{\sqrt{\mu\epsilon}} \qquad f_0 =$ cut-off frequency wavelength in the guide, and f_0 is the cut-off frequency in cycles per second.

The $TE_{1,1}$ mode is known as the "dominant mode" in a cylindrical guide, as it has the lowest cut-off frequency for a given size of guide. It has a general similarity to the "dominant" $TE_{0,1}$ mode in a rectangular guide; indeed, if a $TE_{0,1}$ wave in a rectangular guide passes into a cylindrical guide, it appears there as the $TE_{1,1}$ mode. Note that there is no particular similarity between the $TE_{1,1}$ and the $TM_{1,1}$ cylindrical modes either in appearance or in desirable characteristics.

In a cylindrical guide, the first subscript used to identify the mode is the order of Bessel function in the axial component, and it is also the coefficient of θ in the trigonometric factor. The second subscript is, for a TM mode, the number of the root of the Bessel function at the guide wall (where $r = a$); and for a TE mode it is the number of the root of the derivative of the Bessel function at the guide wall. It is necessary to have a root of the appropriate function at the wall of the guide in order to have zero tangential electric field at the wall.

In any pattern, as in Fig. 80, the first subscript can be determined by counting the number of zero values of E_r as one passes around the circumference of the guide, and dividing by 2. The second subscript is the number of zero values of E_θ along a radius, counting the zero at the wall but not a zero that may exist at the center (or, if E_θ does not exist, use E_z).

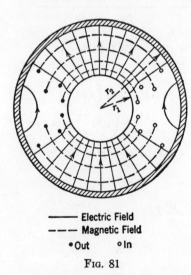

——— Electric Field
– – – Magnetic Field
• Out   ○ In

Fig. 81

Modes in Coaxial Lines. A coaxial transmission line may carry modes other than the simple TEM pattern of radial electric field and circumferential magnetic field. The chief interest in these higher modes lies in knowing how to avoid them. This is usually done by using a coaxial line too small to propagate any higher mode.

Both TM and TE modes are possible between the two conductors of a transmission line. The mode on a coaxial line with the lowest cut-off frequency, and therefore the one to be avoided in the choice of dimensions, has a transverse electric field with the electric field directed outward in half the space and inward in the other half. See Fig. 81. The magnetic field has an axial component. This mode is very like two rectangular $TE_{0,1}$ modes, as in Fig. 74, wrapped around the inner conductor of the transmission line.

As the cut-off wavelength of the rectangular $TE_{0,1}$ mode is $2b$ (equation 424), it is to be expected that the free-space wavelength of the lowest frequency propagated as in Fig. 81 would be about equal to the average circumference of the coaxial line. This turns out to be true, and it is a good approximation that

$$\lambda_0 = 2\pi \frac{r_2 + r_1}{2} \qquad [465]$$

(This is correct within 4 per cent, if $r_2/r_1 < 5$.)

Sending and Receiving Guided Waves. A wave is launched into a wave guide by producing in one end of the guide an electromagnetic field that resembles the field of the desired mode. Thus waves may enter a guide from a cavity resonator, the mode being determined by the field pattern in the cavity at the mouth of the guide.

If the inner conductor of a coaxial transmission line is abruptly terminated, the outer conductor, continuing on, may act as a wave guide. A $TM_{0,1}$ wave will propagate beyond the end of the inner conductor if the outer conductor is large enough to act as a wave guide. Similarly, a $TM_{1,1}$ wave will come off the end of a two-wire line into a cylindrical guide, as might be guessed from Fig. 80. In a rectangular guide, a $TM_{1,1}$ wave would be launched from a coaxial line.

Little antennas in the form of probes or loops within the guide are used with a good deal of ingenuity to produce an approximation to the desired field pattern. Thus, a straight wire entering the side of the guide and projecting parallel to the E lines will launch a $TE_{1,1}$ wave in a circular guide or a $TE_{0,1}$ wave in a rectangular guide. A loop entering the end of the guide in an axial plane will serve the same purpose. A loop in a transverse plane will produce a $TE_{0,1}$ mode in a cylindrical guide.

Transmission in the wrong direction is prevented by closing the guide at an appropriate distance behind the antenna. A metal plate will serve to reflect energy back toward the antenna, reinforcing the signal that travels down the guide in the right direction.

The electromagnetic field at the point of launching is highly complicated, of course; it may be considered as the summation of many superimposed modes. The higher modes will fail to be propagated. Commonly the guide is so designed that only one desired mode will travel along the guide, others being beyond cut-off.

Any device used for starting a wave is equally effective for receiving at the other end of the guide. Energy may thus be delivered from a guide to a line, or into a cavity.

If the end of the guide is merely left open, energy will radiate into space. Radiation from an abruptly terminated guide is not very efficient, for the fields of the guided wave cannot pass directly into being the fields of an unbounded wave in free space. They do not match, and much of the energy is reflected back into the guide. Standing waves result in the guide, as in the case of any mismatch. But, if the end of the guide is flared as a horn, it provides a gradual transition from guide to space. Thus, radiation from a horn may be made quite effective. Some modes, with electromagnetic patterns approximating those of waves in unbounded space, radiate more freely than others. A well-designed horn is a practical radiator and may have good directional characteristics.

Distortion. The previous discussion has been limited to the propagation in wave guides of sinusoidal waves of a single frequency. Modulated waves require the presence of more than one frequency—of side bands as well as carrier. The conclusions that have been reached are not significantly altered if the band of transmitted frequencies is narrow. However, if the band of frequencies is wide, as in the short square pulses of radar and television, it may be necessary to take into account the propagation characteristics of the wave guide. Any time function may be analyzed into sinusoidal components by Fourier analysis. The fundamental and various harmonics (being of different frequencies) will travel at somewhat different speeds along the guide. The fundamental and harmonic components will therefore not have the same phase relation at the receiving end that they have at the sending end. The total wave will be different in form. The general tendency will be for corners to be rounded and other distinguishing features of the wave to become blurred.

Effect of Loss. It is usually safe to neglect energy loss and the attenuation of guided waves in short guides. However, in long guides, or in guides with solid or liquid dielectric material, the loss may be considerable. Further, loss is high and attenuation is rapid in any guide if the transmitted frequency is only slightly above the cut-off frequency.

Let us first consider loss in the metal walls of a guide. The previous discussion of wave guides has neglected loss. If it is necessary to take conduction loss into account, it is usually done by assuming the accuracy of the foregoing results for all guide characteristics except attenuation and then computing attenuation from the loss of energy in the guide walls. This method is not strictly correct, for the presence of loss introduces some small change in all characteristics. Because of the resistance of the guide walls, the flow of current is never quite so great as it should be, the wave is slightly bent back at the edges, and its rate of propagation is delayed. Its wavelength is changed. The Poynting vector field of the wave is so directed that it has a small component into the metal walls. With moderate amounts of attenuation, however, all these effects are so small as to be negligible. This is fortunate, for an exact solution of the wave equation in a guide with walls of finite conductivity is difficult.

It may be generally stated that all walls carry current, for if they did not they might be removed without affecting the operation of the guide. To determine current distribution, let us consider that the purpose of the guide is to permit current to flow in the walls and thereby to provide boundaries for the magnetic field of the transmitted wave. It is known from Chapter X that a metal surface may act as a boundary for an alternating magnetic field, and that, when it does so, the magnetic field is tangential to the metal surface. The direction of current in the metal surface is normal to the direction of the tangential magnetic field (for it is in the direction of the curl of the field), and the amount of current per unit length of guide surface (by equation 320) is numerically equal to the tangential component of **H**. Hence, by considering a diagram of magnetic field, the current in the guide walls may be mapped: lines of current flow are normal to the magnetic flux lines at the surface, and they are most dense where the magnetic field is most dense.

It is not necessary, in computing current, to take into account the electric field that terminates on walls of the guide. It was shown early in this chapter that the same current may serve both to terminate the electric field and at the same time to provide the correct boundary for the magnetic field. This relation is inherent in any correct solution of Maxwell's equations. Hence, whether there is or is not electric field terminating on the wall of a guide, the current in the wall can be found from the tangential magnetic field alone.

Having found the current in the wall, loss is determined from the "surface resistivity" which was defined in Chapter X. The surface resistivity is a function of frequency, being in reality a skin-effect phenomenon. It is equal to $1/\gamma\delta$, and since δ is the effective depth of penetration of current the surface resistivity may also be written

$$R_s = \sqrt{\frac{\pi f \mu}{\gamma}}$$

For silver: $\gamma = 6.14 \times 10^7$; $R_s = 2.53 \times 10^{-7}\sqrt{f}$
For copper: $\gamma = 5.80 \times 10^7$; $R_s = 2.61 \times 10^{-7}\sqrt{f}$
For aluminum: $\gamma = 3.54 \times 10^7$; $R_s = 3.34 \times 10^{-7}\sqrt{f}$

It is important to recognize that the surface resistivity is directly proportional to the square root of frequency; this corresponds to a decreased "skin depth" at high frequency. Loss is found as $I^2 R_s$.

To find attenuation after loss is known, it is necessary to compute the total energy transmitted along the guide. To do this, the Poynting vector field $\mathbf{E} \times \mathbf{H}$ is integrated through any cross section of the guide and averaged with respect to time; the result is power at that section of the guide. Attenuation is then found in terms of power transmitted and power lost. In any length of guide

$$P_{\text{out}} = P_{\text{in}} e^{-2\alpha x}$$

(Both the electric and magnetic fields are attenuated at the same rate α, so power is attenuated at the rate 2α.) Since $P_{\text{loss}} = P_{\text{out}} - P_{\text{in}}$, we have, for a 1-meter length of guide:

$$\frac{P_{\text{loss}}}{P_{\text{in}}} = 1 - e^{-2\alpha}$$

Expanding the exponential in series and retaining only the first two terms gives an approximation good for small attenutation:

$$\alpha = \frac{P_{\text{loss}}}{2P_{\text{in}}}$$

A general solution for loss in all guides will not be attempted, but the main steps in finding attenuation in a rectangular guide carrying a $TE_{0,1}$ mode are indicated to illustrate the general principles. From Table VI:

$$E_y = \frac{A_0}{b} \sin \frac{\pi z}{b} e^{j\omega t - \Gamma x}$$

$$H_x = \frac{A_0}{j2f\mu b^2} \cos \frac{\pi z}{b} e^{j\omega t - \Gamma x}$$

$$H_z = \frac{A_0 G}{\eta b} \sin \frac{\pi z}{b} e^{j\omega t - \Gamma x}$$

In the bottom of the guide (Fig. 70) the square of the magnitude of surface current is

$$|I|^2 = |I_x|^2 + |I_z|^2 = |H_z|^2 + |H_x|^2$$

Multiplying I^2 by R_s gives an expression for power loss per unit area. The above values for H_x and H_z are then substituted into the expression, and the result is loss per unit area at a given cross section ($x = $ constant) as a function of z and t. The width z appears as a sine or cosine squared function, the average of which is one-half the maximum. Time appears as the square of an exponential function, and, considering that only the real component of the exponential is significant, its average is also one-half the maximum. The time average of loss in unit length of both top and bottom surfaces of the guide of width b is then

$$2R_s b \left(\frac{A_0^2 G^2}{4\eta^2 b^2} + \frac{A_0^2}{16 f^2 \mu^2 b^4} \right)$$

The average loss per unit length of both the vertical walls ($z = 0$) is similarly found from current distribution in those walls to be:

$$2R_s a \frac{A_0^2}{8 f^2 \mu^2 b^4}$$

Average transmitted power is the integral of the Poynting field, or, in this case, the average of the product of E_y and H_z times the cross-section area:

$$\frac{A_0^2 G}{4 b^2 \eta} ab$$

Attenuation (nepers per meter) owing to imperfect conductors is then half the ratio of power loss to power transmitted, or

$$\alpha = \frac{R_s \left[1 + \frac{2a}{b} (1 - G^2) \right]}{a \eta G}$$

Attenuation in decibels per meter, the value usually desired for practical work, is 8.686α.

If the transmitted frequency is high above the cut-off frequency, the guide factor G approaches one, and α is approximately $R_s/a\eta$, increasing as the square root of frequency. If the transmitted frequency is near the cut-off frequency, G approaches zero, and the attenuation rises without limit. At an intermediate frequency, some two or three times cut-off frequency, there is minimum attenuation. Curves of attenuation for the rectangular $TE_{0,1}$ mode are shown in Fig. 82, and these are fairly typical of other modes and guides. The reason for great attenuation near cut-off is that the axial component of field becomes relatively strong (as G approaches zero) and contributes to the loss without contributing to the transmission of energy. Increase of attenuation at very high frequency, on the other hand, results from extreme skin effect.

One of the most remarkable waves is the $TE_{0,1}$ mode in a circular guide. In this particular wave, shown in Fig. 80, all electric flux lines as well as all magnetic flux lines are closed loops. None of the electric

224   WAVE GUIDES

field terminates on the wall of the guide. The only current carried by the wall is to provide a boundary for the axial component of magnetic field. Since, in this mode (see Table IX) as in all others, the axial component of field becomes relatively less at high frequency, the current in the walls becomes less, and this mode has the unique characteristic of being less attenuated as the frequency is increased indefinitely. For this mode the curves of Fig. 82 are not typical.

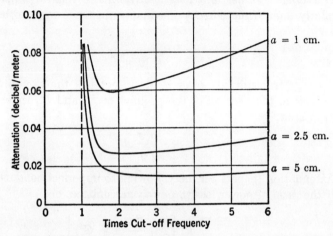

FIG. 82. Attenuation of $TE_{0,1}$ mode in a rectangular copper wave guide; $b = 5$ cm.

If there is loss in the dielectric medium within a wave guide, the resulting attenuation is an additional factor to be determined. The imperfection of the dielectric can be included in the wave-guide equations in the same way it was included in the equations for unbounded waves in Chapter IX. When conductivity of the dielectric is not zero, the wave equation, derived from Maxwell's equations, must include γ. Assuming a sinusoidal function of time we write, instead of equation 435,

$$\nabla^2 \mathbf{E}_0 = (-\omega^2 \mu \epsilon + j\omega\mu\gamma)\mathbf{E}_0$$

The wave-guide solution continues from this equation, paralleling the solution for waves in the lossless guide, except that we find

$$\Gamma^2 = -\omega^2 \mu \epsilon \left(G^2 - \frac{j\gamma}{\omega\epsilon} \right)$$

The real part of Γ is α. We introduce (as in Chapter IX) the power factor of the dielectric as the most convenient practical measure of loss and from a binomial expansion of Γ there results, as a good approximation of attenuation owing to dielectric loss,

$$\alpha = \frac{\omega\sqrt{\mu\epsilon}}{2G} \times \text{(power factor)}$$

This gives α in nepers per meter for any mode or guide. For decibels per meter, multiply by 8.686.

Attenuation owing both to imperfect conduction of guide walls and to imperfect dielectric material is the sum of the two component attenuations. The α's may be added either as nepers or as decibels per meter.

Guides and Lines. Current in the cylindrical guide for the $TE_{0,1}$ wave circulates around the guide. There is no axial component of current along the guide. This seems queer to one accustomed to thinking of a transmission line as a means of conveying current. But when we consider that current flows in a guide or line only for the purpose of providing a boundary for the electric and magnetic fields, a new concept results.

From this point of view, the conductors of all wave guides and transmission lines are for the purpose of carrying charge and current to terminate the fields and to permit the existence of waves which, being guided, do not spread their energy uselessly through space. This concept can be extended, indeed, to all electric wiring, the purpose of which is to guide energy in the magnetic and electric fields—current being incidental.

Although a conducting surface is the best wave guide, it is not the only possible kind. A dielectric surface will also serve as a boundary for certain types of waves. The discontinuity between material of high dielectric constant and low dielectric constant makes it possible to confine a wave within the material of high constant. The practical objection to the use of a dielectric wave guide is that the loss in all known dielectric materials is too great for satisfactory wave transmission.

The most suitable wave guide for any particular application depends on the conditions and particularly the frequency. At power frequency, the parallel-wire transmission line is the best guide. At high radio frequency, the concentric-conductor line is more desirable, for, although more expensive to build, it has negligible radiation loss. Hollow wave guides are easier to construct and to use at the frequencies of centimeter and millimeter waves, and they work well with certain types of tubes, cavity resonators, horns, and highly directional radiators. There is no doubt that all three types of wave guides will continue to be used.

PROBLEMS

1. A wave following a concentric transmission line, in which one conductor is a solid cylinder and the other a coaxial hollow cylinder, may be described in the empty space between the conductors as

$$E_r = \frac{E_0}{r} \sin \frac{\omega}{v} (z - vt)$$

$$E_\theta = E_z = 0$$

(a) This wave can exist only if the electric field has no divergence, and if it is a solution of the wave equation, equation 435. Determine whether these conditions are satisfied.

(b) From Maxwell's equations, determine the magnetic field of this wave.

(c) Find the current in the inner conductor, assuming perfect conductivity. Show that this current provides a proper boundary for both the electric and magnetic fields.

(d) Find the Poynting vector field.

(e) Find the velocity of the wave if the space between conductors is filled with oil, as in a power cable, of dielectric constant 2.17.

2. Relate the group velocity of a wave in the hollow rectangular guide of Fig. 60 to the angle between the paths of the elementary wave components and the axis of the guide. Call this angle α.

3. Find the angle α of Problem 2 in terms of the wavelength of the elementary wave component and the dimension b of the guide (see Fig. 75). From this derive equation 426.

4. Find the phase velocity from Fig. 76, and show that it may be expressed as equation 427.

5. A wave enters a wave guide with the form $E = E_m(\sin \omega t + \frac{1}{3} \sin 3\omega t)$. Plot the shape of the wave as it enters, and as it passes various positions along the guide. The cut-off frequency of the guide is half the frequency of the fundamental component of this voltage. Neglect attenuation.

6. Show that the divergence of the electric field described by equation 429 is zero in the space within the guide.

7. (a) Show that the divergence of the magnetic field of equation 434 is zero.

(b) Prove that the divergence of *any* magnetic field given by Maxwell's equation 193 will be zero.

8. In footnote 1, page 195, $\nabla \cdot \iota = - \partial\rho/\partial t$ is derived from equations 240 and 115. Show that this expression can be deduced, instead, from the definition of current density in terms of charge and can then be combined with equation 240 to derive equation 115 for all non-static fields.

9. Derive equation 452 from equation 435, paralleling the derivation of equation 451 in the text.

10. Express K_2 and K_3 in equations 456 in terms of the constants of equations 452 and 453.

11. Derive equations 459, using Maxwell's equation as suggested.

12. A hollow guide is 5 centimeters square. Plot the factor G (Table VIII) for the $TE_{1,0}$ mode for frequencies from zero to cut-off, and from cut-off to 5 times cut-off. Compute the cut-off frequency.

PROBLEMS

13. Show from the general equations that the rectangular $TM_{1,0}$ and $TM_{0,1}$ modes cannot exist.

14. A hollow guide is 5 centimeters square. What is the cut-off frequency for each of the following modes: $TE_{1,0}$; $TE_{1,1}$; $TM_{1,1}$; $TM_{1,2}$?

15. How does the rectangular $TE_{m,n}$ mode differ from the $TE_{n,m}$ mode?

16. Sketch the $TE_{1,2}$ mode in a square guide, as other modes are shown in Fig. 79.

17. Sketch the vector fields of current density in the guide walls for the following rectangular modes: $TE_{0,1}$; $TE_{1,1}$; $TM_{1,1}$. Indicate the relative positions of current density and electric and magnetic fields. Because of symmetry, the currents in any two perpendicular walls will be enough for each mode.

18. A hollow guide is 5 centimeters square. The $TE_{1,0}$ mode is excited in the guide at frequencies as follows: (a) 90 per cent of cut-off; (b) 80 per cent of cut-off; (c) half cut-off. At what distance along the guide is the signal strength reduced to 1 per cent of the excitation?

19. Using tables of Bessel functions, show that the tangential electric field is zero at the wall of the guide in the four cylindrical modes of Table IX.

20. A hollow cylindrical guide is 5 centimeters in diameter. Find the cut-off frequency for each of the four modes of Table IX. Compare with results of Problem 14.

21. Compare the meaning of G in Table VIII and Table IX.

22. The signal frequency in a 5-centimeter square hollow copper guide is 4000 megacycles per second. It happens that both the $TE_{1,0}$ and the $TE_{1,1}$ modes are excited. Compare the "reactive" attenuation of the $TE_{1,1}$ mode with the dissipation attenuation of the $TE_{1,0}$ mode which may be approximated from Fig. 82.

23. The guide of Problem 22 is filled with polystyrene for which (in this frequency range) $\kappa = 2.50$ and the power factor is 0.0008. What is the additional attenuation of the 4000-megacycle signal in the $TE_{0,1}$ mode resulting from dielectric loss? Compare to this the attenuation resulting from guide resistance, as found in Problem 22, noting however that the latter is only an approximation for the dielectric-filled guide.

CHAPTER XIV

Waves in the Ionosphere

Group Velocity and Phase Velocity. The difference between group and phase velocity in a wave guide can be explained in terms of elementary waves traveling with continual reflections. The wave guide is a special case, however, and it is not usually possible to account for group and phase velocity on the basis of component waves. The *general* condition is this: If the velocity of propagation of waves depends upon the wavelength, group velocity will differ from phase velocity.

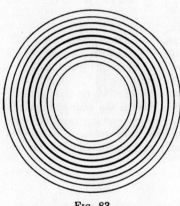

Fig. 83

The most familiar illustration is found in waves that spread over the surface of water. If a stone is dropped into a quiet pond, a band of concentric circular ripples will travel outward from the point of disturbance. There will be a number of waves in the group, as in Fig. 83: the waves near the middle of the group will have the greatest amplitude, and the inner and outer waves will be vanishingly small. If one watches the waves with care, the individual waves will be seen to travel faster than the group as a whole. A wave will appear at the inner circumference of the band and will gain amplitude as it moves outward, while other waves appear, one by one, behind it. After the wave has passed the middle of the band of waves, however, it will diminish in amplitude, until it becomes the outermost wave and finally vanishes. The individual wave moves with phase velocity; the band of waves moves with group velocity.

In Fig. 84, a cross section of such a band of waves is shown. It is indicated in the diagram that the band can be considered as the sum of two waves of constant amplitude but slightly different frequency (This obviously corresponds to the analysis of a modulated radio wave

GROUP VELOCITY AND PHASE VELOCITY

into carrier frequency and side bands.) Two of the crests of the wave of shorter wavelength are marked 1 and 2; two crests of the wave of greater wavelength are marked a and b. The crests 1 and a coincide at the instant for which the diagram is drawn, and they add to give the greatest of the crests in the resultant wave, marked I. The position of the group of waves (that is, the position of the dotted envelope) is given by the position of this maximum crest.

Now assume that the wave in Fig. 84 is traveling from left to right, and that the nature of the medium is such that long waves travel a little faster (phase velocity) than short waves. At a slightly later

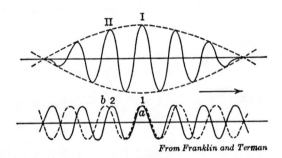

From Franklin and Terman

FIG. 84

instant of time, crest b will overtake crest 2. When this happens, crest a will have moved on beyond crest 1. Consequently, at this later instant (all waves having moved a considerable distance toward the right), the crest marked II will have become the maximum and central crest of the resultant group. Thus, the center of the group of waves will have moved a lesser distance to the right (less by one wavelength) than the component waves of Fig. 84. The speed of the group of waves, known as group velocity, is therefore less (in this instance) than phase velocity.

Mathematically, the phase velocity of a wave is

$$v_\phi = \frac{\omega}{\beta} \qquad [466]$$

as in Table III. The group velocity is

$$v_g = \frac{d\omega}{d\beta} \qquad [467]$$

Therefore if β, the phase constant, can be expressed as a function of

WAVES IN THE IONOSPHERE

frequency and differentiated, the group velocity can be obtained.[1] By this means, equation 426 which gives group velocity in a wave guide can be obtained from phase velocity in equation 427 or 432.

Application of this formula to water waves shows that the phase velocity is twice the group velocity (if the waves are large enough to be independent of surface tension), as follows.

The phase velocity of water waves is known to be

$$v_\phi = \frac{g}{2\pi f} = \frac{g}{\omega} \qquad [472]$$

where g is acceleration due to gravity and f is frequency of the wave. It will be seen that high-frequency waves are propagated more slowly. From equation 466

$$\beta = \frac{\omega}{v_\phi} = \frac{\omega^2}{g} \qquad [473]$$

and differentiation gives

$$\frac{d\beta}{d\omega} = \frac{2\omega}{g}$$

whence

$$v_g = \frac{d\omega}{d\beta} = \frac{g}{2\omega} = \frac{1}{2} v_\phi \qquad [474]$$

as stated above.

[1] Derivation of equation 467 follows. Describe the two waves of Fig. 84 as

$$A = M\, e^{j[(\omega - \Delta\omega)t - (\beta - \Delta\beta)x]} + M\, e^{j[(\omega + \Delta\omega)t - (\beta + \Delta\beta)x]} \qquad [468]$$

This defines a wave of two components of equal amplitude, but differing slightly in frequency and in the value of β. Rearranging terms:

$$A = M\, [e^{j(\Delta\omega t - \Delta\beta x)} + e^{-j(\Delta\omega t - \Delta\beta x)}]\, e^{j(\omega t - \beta x)} \qquad [469]$$

Recognizing the quantity in brackets as twice the cosine, and identifying the real part of the final exponential as a cosine:

$$A = 2M \cos(\Delta\omega t - \Delta\beta x) \cos(\omega t - \beta x) \qquad [470]$$

This is the product of a traveling wave of low frequency (the envelope) and a traveling wave of high frequency (the carrier). The carrier frequency is the average of the frequencies of the two component waves. The velocity of this high-frequency carrier wave, which is phase velocity, is ω/β. The envelope frequency is half the difference between the frequencies of the two component waves. The velocity of the envelope is $\Delta\omega/\Delta\beta$. The group velocity is the limit of the envelope velocity as $\Delta\omega$ becomes small:

$$v_g = \lim_{\Delta\omega \to 0} \frac{\Delta\omega}{\Delta\beta} = \frac{d\omega}{d\beta} \qquad [471]$$

It must be understood that, when the wave motion is sinusoidal and steady, group velocity is not apparent. There are no groups; only individual waves can be observed, and they travel with phase velocity. If two frequencies are present, as in the illustration of Fig. 84, or three frequencies, such as the carrier and two side frequencies of a steadily modulated wave, the situation is fairly simple. However, in a transient disturbance, such as the ring of waves that results from dropping a stone into water or the short section of radio wave released as a telegraphic dot or dash, all frequencies are present. The result is then exceedingly difficult to analyze,[2] although it is qualitatively similar to the simple case.

To the electrical engineer, the most familiar variation of velocity with frequency is on a transmission line with losses. If the resistance and leakage losses are not proportioned to give the "distortionless" condition, phase velocity of waves will be less than on the same line without losses. The phase velocity of low-frequency waves will be less than the phase velocity of high-frequency waves. Group velocity can be computed from phase velocity. On an ordinary transmission line, group velocity will always be *greater* than phase velocity.[3]

Another example of variation of phase velocity with frequency is the propagation of light through transparent material. In glass, for example, high frequencies (blue light) travel more slowly than low frequencies (red light). One result of this action is the well-known dispersive effect of prisms. Phase and group velocities are both less than the speed of light in free space, and except in special cases the group velocity is less than the phase velocity.

Phase velocity of radio waves in the ionosphere is dependent upon frequency. The effect, in this case, results from vibratory motion of electrons in the ionized layer and reradiation of energy in different phase. The net result is a phase velocity greater than the velocity of light, and group velocity less than the velocity of light. The group velocity can be quite low, and this is believed by some to account for the occasional observation of signals that reach a radio receiving station as much as several seconds later than they would have arrived if they had been propagated at the speed of light in free space.

The Ionosphere. The ionosphere is a region above the surface of the earth in which there are considerable numbers of free electrons and ionized molecules. The ionization exists in fairly well-defined layers at about 100 kilometers, 200 kilometers, and higher, varying between

---

[2] See, for instance, *Communication Networks*, Volume II, E. A. Guillemin, John Wiley & Sons, New York, 1935.
[3] See E. A. Guillemin, *loc. cit.*

day and night and from season to season. Radiation from the sun is believed to be the cause of ionization. Radio waves reaching the ionosphere are reflected back to the surface of the earth if they do not exceed a critical frequency. The critical frequency for a wave approaching the ionosphere obliquely is greater than that for a wave rising vertically from the earth and approaching the ionosphere at normal incidence. There are no precise boundaries to the ionized layers; the ionization increases from a negligible value to maximum through a distance of kilometers; radio waves therefore penetrate more or less deeply into the ionosphere and are affected by the presence of ionization, even though they are finally returned to earth.

The ionosphere does not reflect electric waves in the same manner as a metallic conductor because, first, the number of free electrons in the ionosphere is relatively limited, and, second, the ionosphere electrons are comparatively free to move without interference. An important factor in the behavior of a wave in an ionized region is the number of electrons per cubic meter. (Ions other than electrons might be considered in the same analysis; practically, however, experimental data show that most of the effect of the ionosphere is the result of free electrons, and this is to be expected because of their much greater mobility.)

When a radio wave passes an electron, the electric field of the wave exerts force on the charged electron, as on any charged body. The electron moves in response to the force, and as the electric field alternates during the passage of the wave the electron vibrates. Motion of the electron constitutes current, which in turn affects the propagation of the wave.

If the vibrating electron collides with molecules of air it wastes energy, but if it is free to vibrate without collisions there will be no energy dissipation. In the higher levels of the ionosphere and at high radio frequency, an electron may vibrate many times without a collision. In lower levels where the air is denser and at frequencies as low as standard broadcast frequency, loss may be quite important in attenuating the wave.

Since a moving electron is an electric current, it will have force exerted on it by magnetic fields. The effect of the magnetic component of the passing wave is too slight to require consideration, but the earth's magnetic field has an appreciable effect; the main result of the earth's field is to introduce new components of polarization into the wave.

If all effects are considered, the behavior of a wave in the ionosphere is rather complicated. The basic principles of ionosphere action are best shown by making certain simplifying assumptions. Fortunately, the assumptions to be made are very good approximations for a wide range of practical operation.

Waves in an Ionized Region. Let us consider an ionized region in which there is so little air that electrons may vibrate without colliding with molecules, and let us neglect the earth's magnetic field. An electric wave is passing through the region; it is sinusoidal:

$$\mathbf{E} = \mathbf{E}_0\, e^{j\omega t} \qquad [475]$$

The mass of an electron is m and its charge is q (the more usual symbol e is used with another meaning). Mass times acceleration equals force, and, if the $\mathbf{E}$ is polarized along the X axis,

$$m\frac{d^2x}{dt^2} = qE_x \qquad [476]$$

Substituting equation 475 into 476 and integrating once,

$$m\frac{d^2x}{dt^2} = qE_{0x}\, e^{j\omega t} \qquad [477]$$

$$\frac{dx}{dt} = \frac{q}{j\omega m}E_x \qquad [478]$$

This first derivative dx/dt is velocity of electronic motion. If there are N electrons per cubic meter, each having a charge of q and each moving with the velocity of equation 478, they provide a current density of

$$\iota_x = Nq\frac{q}{j\omega m}E_x \qquad [479]$$

More generally,

$$\iota = -j\frac{Nq^2}{\omega m}\mathbf{E} \qquad [480]$$

This electron current is out of phase with the electric field that produces it, and this is the chief difference between current that flows in the ionosphere in response to a passing wave, and current in any solid conductor. In a metal, motion of the electrons is limited by the resistance of the material; in the ionosphere, with the air density so low that collisions with molecules can be neglected, motion is limited only by the inertia of the electrons, and the velocity of the electrons is consequently out of phase with the driving force.

Various short-cuts are possible in deriving equations of wave propagation in an ionized medium, but it seems worth while to go all the way back to Maxwell's equations as a logical starting point. Introducing equation 480 into Maxwell's equations:

$$\nabla \times \mathbf{H} = \iota + \frac{\partial \mathbf{D}}{\partial t} = -\frac{jNq^2}{\omega m}\mathbf{E} + \epsilon \frac{\partial \mathbf{E}}{\partial t} \qquad [481]$$

$$\nabla \times \mathbf{E} = -\frac{\partial \mathbf{B}}{\partial t} = -\mu \frac{\partial \mathbf{H}}{\partial t} \qquad [482]$$

We assumed a sinusoidal electric field in equation 475. Since relations between the electric and magnetic fields and current are all linear, it follows that $\mathbf{H}$ will be sinusoidal, also:

$$\mathbf{H} = \mathbf{H}_0 \, e^{j\omega t} \qquad [483]$$

These expressions in Maxwell's equations give

$$\nabla \times \mathbf{H}_0 = \left(-\frac{jNq^2}{\omega m} + j\omega\epsilon\right)\mathbf{E}_0 \qquad [484]$$

$$\nabla \times \mathbf{E}_0 = -j\omega\mu \mathbf{H}_0 \qquad [485]$$

Solving as usual, we take the curl of equation 484. The identity $\nabla \times \nabla \times \mathbf{H} = \nabla(\nabla \cdot \mathbf{H}) - \nabla^2 \mathbf{H}$ is used, noting that the divergence of the magnetic field is zero. Thus

$$\nabla \times \nabla \times \mathbf{H}_0 = -\nabla^2 \mathbf{H}_0 = \left(-\frac{jNq^2}{\omega m} + j\omega\epsilon\right)(\nabla \times \mathbf{E}_0) \qquad [486]$$

Using equation 485:

$$\nabla^2 \mathbf{H}_0 = -\omega^2 \mu\epsilon \left(1 - \frac{Nq^2}{\omega^2 \epsilon m}\right) \mathbf{H}_0 \qquad [487]$$

For convenience we shall write

$$\Gamma = \omega \sqrt{-\mu\epsilon\left(1 - \frac{Nq^2}{\omega^2 \epsilon m}\right)} \qquad [488]$$

and, when this symbol is used, equation 487 becomes

$$\nabla^2 \mathbf{H}_0 = \Gamma^2 \mathbf{H}_0 \qquad [489]$$

which is the familiar wave equation.

Let us discuss a specific wave. We may select a polarized plane wave defined by

$$H_y = H_{0y} \, e^{j\omega t} \qquad H_x = H_z = 0 \qquad [490]$$

$$\frac{\partial H_y}{\partial x} = \frac{\partial H_y}{\partial y} = 0 \qquad [491]$$

WAVES IN AN IONIZED REGION

See Fig. 85. When this wave is used in equation 489, it becomes:

$$\frac{\partial^2 H_{0y}}{\partial z^2} = \Gamma^2 H_{0y} \qquad [492]$$

The solution of this simple differential equation may [4] be

$$H_{0y} = H_m e^{-\Gamma z} \quad \text{or} \quad H_y = H_m e^{j\omega t - \Gamma z} \qquad [493]$$

Now we find the curl of **H**, and from equation 484 we write:

$$-\frac{\partial H_y}{\partial z} = \Gamma H_y = \frac{\Gamma^2}{j\omega \mu} E_x \qquad [494]$$

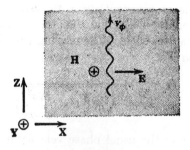

⊕ : Vector into Page

Fig. 85

This tells us that E_x is a wave similar to H_y, and may be written

$$E_x = E_m e^{j\omega t - \Gamma z} \qquad [495]$$

Also, if the intrinsic impedance is defined by $E_x = \eta H_y$, we have from equation 494,

$$\eta = \frac{j\omega \mu}{\Gamma} = \sqrt{\frac{\mu}{\epsilon \left(1 - \frac{Nq^2}{\omega^2 \epsilon m}\right)}} \qquad [496]$$

It will be recognized from the wave equation that Γ is the propagation factor for a wave in an ionized region. It is either real or imaginary; it cannot be complex. If it is imaginary, the wave is undamped and travels with undiminished amplitude. Writing $\Gamma = j\beta$, the phase constant is

$$\beta = \omega \sqrt{\mu \epsilon \left(1 - \frac{Nq^2}{\omega^2 \epsilon m}\right)} \qquad [497]$$

[4] The sign of Γ may equally well be positive in 493, giving a wave traveling in the opposite direction.

If the density of ionization is vanishingly small, the phase constant reduces to $\beta = \omega\sqrt{\mu\epsilon}$, as in a perfect dielectric medium. For larger values of N, β is less, and a critical value of N will make β zero.

If, on the contrary, Γ is real, there will be no wave propagation in the ionized medium at all, for equation 495 does not describe a wave unless Γ has at least a component that is imaginary. With a real value of Γ, equation 495 describes a pulsation that is everywhere in phase, diminishing with positive values of z. This implies that, if a region is so heavily ionized and the frequency of an electric disturbance is so low that $\dfrac{Nq^2}{\omega^2\epsilon m} > 1$, the disturbance will produce not a wave but merely a local field that vanishes exponentially with distance.

Velocity. It is remarkable to find that the vibration of electrons in an ionized region has the effect of increasing the phase velocity. We have, from equation 497,

$$v_\phi = \frac{\omega}{\beta} = \frac{1}{\sqrt{\mu\epsilon\left(1 - \dfrac{Nq^2}{\omega^2\epsilon m}\right)}} \qquad [498]$$

If N is zero, this gives the usual phase velocity in a dielectric. With increasing values of N, the velocity is higher, and, as the critical value is approached beyond which wave propagation is impossible, the phase velocity approaches infinity.

Group velocity of such a wave, however, is less than in a perfect dielectric. Knowing that $v_g = d\omega/d\beta$, the group velocity is easily found. From equation 497

$$\frac{d\beta}{d\omega} = \frac{\mu\epsilon}{\sqrt{\mu\epsilon\left(1 - \dfrac{Nq^2}{\omega^2\epsilon m}\right)}} \qquad [499]$$

whence

$$v_g = \frac{d\omega}{d\beta} = \frac{1}{\sqrt{\mu\epsilon}}\sqrt{1 - \frac{Nq^2}{\omega^2\epsilon m}} \qquad [500]$$

Since any ionized region in which electrons vibrate freely can hardly be imagined to have permeability or dielectric constant differing appreciably from free space,[5] μ and ϵ in equations 498 and 500 might well be written

[5] Perhaps it should be emphasized that μ and ϵ have their ordinary meanings in this derivation. Some authors assign a fictitious value of $\epsilon\left(1 - \dfrac{Nq^2}{\omega^2\epsilon m}\right)$ to the dielectric constant of an ionized region in order to take into account the action of the electrons; such a value is less than ϵ, which accounts for the higher phase velocity.

μ_0 and ϵ_0. This is certainly true in the ionosphere. Remembering that the velocity of a plane wave in free space is $c = 1/\sqrt{\mu_0 \epsilon_0}$, phase and group velocities can be written:

$$v_\phi = \frac{c}{\sqrt{1 - \dfrac{Nq^2}{\omega^2 \epsilon_0 m}}}$$

$$v_g = c\sqrt{1 - \frac{Nq^2}{\omega^2 \epsilon_0 m}}$$

[501]

It is interesting to notice that

$$v_\phi v_g = c^2 \qquad [502]$$

Intrinsic Impedance. The intrinsic impedance of an ionized medium is given in equation 496. This intrinsic impedance is greater than that of an un-ionized region for any value of N less than the critical value, indicating that a wave has a relatively weak magnetic field in comparison with its electric field. For N greater than the critical value, η is imaginary, showing that the electric and magnetic components of the field (not, in this case, a wave) are then out of phase.

Normal Reflection. If a radio wave arrives at a layer of the ionosphere in which ionization is so dense that the wave cannot be propagated, what happens to the energy of the wave? The energy does not pass through, for no wave is transmitted; energy is not absorbed by the ionized layer (neglecting collisions of the vibrating electrons with gas molecules); hence energy must be reflected.

This is, indeed, what happens. If a wave is sent vertically upward from the earth, it approaches an ionosphere layer at normal incidence; it penetrates the lower regions of the layer in which ionization is increasingly more dense, traveling with increasing phase velocity and decreasing group velocity as the height of critical ionization density is approached. Above the critical level there is not wave propagation, but a pulsing field that vanishes at greater heights.

The "virtual height" of the ionosphere is found experimentally by measuring the time required for pulses of waves to travel upward from the surface of the earth to the ionosphere, and down again after reflection. Dividing half the time by the velocity of waves in free space gives the "virtual height." This is not the true height of the level at which N has its critical value but somewhat greater, because the experimental pulses travel with a group velocity less than c while traversing the lower layers of the ionosphere, just before and after reflection.

Virtual height and true height of reflection both depend on frequency; low-frequency waves are reflected from the lowest levels, whereas centimeter waves penetrate all layers of the ionosphere and pass out into inter-planetary space. Radar signals that have been received after reflection from the moon had to be rather high-frequency waves to pass through the earth's ionosphere.

To find whether radio waves will be reflected by the ionosphere if the electron density is known, we may use equation 488. It was seen that upward propagation fails, and hence there is reflection, if

$$\frac{Nq^2}{\omega^2 \epsilon m} > 1 \qquad [503]$$

Using the charge and mass of the electron (1.60×10^{-19} coulomb, and 9.11×10^{-31} kilogram), it is found that there will be reflection of a wave rising *vertically* if the frequency is less than the critical frequency:

$$f = 9.0\sqrt{N} \qquad [504]$$

Frequency [6] is in cycles per second; N is electrons per cubic meter.

Oblique Reflection. Radio waves commonly approach the ionosphere at an oblique angle, rather than at normal incidence. It is usual to discuss the travel of such a wave in terms of refraction, using as index of refraction the ratio of the velocity in free space to the phase velocity in the ionosphere. This is consistent with the usual definition of the index of refraction in optics, and with equation 360 in Chapter X. Thus

$$\text{Index of refraction} = \frac{c}{v_\phi} = \sqrt{1 - \frac{Nq^2}{\omega^2 \epsilon m}} \qquad [505]$$

Then the ionosphere is looked upon as a region of varying index of refraction, in which an oblique wave follows a curved path; the radius of curvature is determined from the rate of change of index of refraction. A ray is turned back to the earth, as in Fig. 86, much as a light wave or a wave in a dielectric guide may be trapped by total reflection, unable to pass obliquely from a region of higher to a region of lower index of refraction. Equation 505 shows that the index of refraction of the ionosphere is less than the index of refraction of free space (which, by definition, is 1).

A simpler mental picture results from considering that phase velocity is greater in the ionosphere. Referring again to Fig. 86, think of the

[6] Note that if N is used as electrons per cubic centimeter, f will be frequency in kilocycles.

upper edge of a wave crest traveling faster than the lower edge because of being in a more highly ionized region. The wave will be turned back to earth again, as shown.

These concepts of "refraction" in the ionosphere are just other ways of looking at the same phenomenon that was also called "reflection."

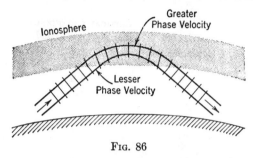

FIG. 86

The phenomenon, whatever it is called, results from transfer of energy of the incident wave to the vibrating electrons of the ionosphere and reradiation of energy by the electrons in such a way as to give a new wave pattern that may, in various cases, be interpreted as reflection or refraction or mere change of speed.

PROBLEMS

1. Derive the group velocity in an ionized region, equation 500, from equation 497. Show steps of differentiation that are omitted in the text.

2. What factor in equations 498 and 500 is analogous to the guide factor G of Chapter XIII? (See Table VIII.) Define such a factor, which may be called G_i and write equations 488, 496, 497, 498, 500, and 501 in terms of G_i. Compare with similar equations in Tables VIII and IX.

3. Assuming that electron density is zero up to some level above the earth and for greater heights is proportional to distance above that level, plot group velocity and phase velocity of a vertically traveling wave as functions of height. Plot from $N = 0$ to the critical value of N.

4. Design a tapered wave guide ($TE_{1,0}$ rectangular mode) to have the same v_ϕ and v_g as functions of distance of travel as were plotted in Problem 3. Will there be reflection in such a guide corresponding to reflection of a vertically incident wave from the ionosphere?

5. A wave in free space suddenly (in a small fraction of a wavelength) enters an ionized region in which N is 0.8 of the critical value. Treating this as a problem of reflection of a wave at normal incidence, find the reflection factor. Explain why this is *not* typical of the ionosphere.

6. If an ionized region contains an equal number of electrons and positively charged hydrogen ions, with equal amount of charge but 1838 times as heavy, what fraction of the total current produced by a passing radio wave will result from vibration of the electrons? Will electron current and ion current add or subtract?

WAVES IN THE IONOSPHERE

7. Following are values of critical frequency, above which there is no normal reflection, as observed at different hours of a winter day at a time of the sun-spot cycle corresponding to relatively intense ionization. These are for reflection of a wave vertically incident on the F_2 layer of the ionosphere. Find and plot the maximum electron density in the reflecting layer throughout the day. Local time is given; sunrise was about 7:30 A.M.

Hour (local time)	Critical Frequency (megacycles/second)
Midnight	5.0
4	4.8
6	4.2
8	8.0
12	11.8
16	9.6
18	7.8
20	6.1
24	5.0

Index

Abraham, M., 164
Accelerator, induction, 104
Advanced potential, 171
Aether, luminiferous, 117
Ampère, André, 77, 110
Ampère's law, 77, 91
Ampere-turn, 86
Amperian current, 96
Antenna, 124, 164
  above ground, 179
  arrays, 186
  equivalent length of, 173
  flat-top, 185
  grounded, 182, 189
  half-wavelength, 174
  heights, 182
  loop, 189
  receiving, 187
  short, 172
Area, 11
  positive, 40, 87
Arrays, antenna, 186
Attenuation, 128, 149, 222
  reactive, 215

Ballantine, Stuart, 192
Barrow, W. L., 200
Becker, Richard, 164
Betatron, 105
Biot-Savart law, 90
Boreal direction, 87
Boundary conditions, 56, 92
Boundary surfaces, 138
Brewster angle, 159, 183

Capacitance, 58, 59
Carson, J. R., 189
Cartesian coordinates, 14, 36
Caterpillar, 203
Cavendish, Henry, 61
Cavity resonator, 219
cgs units, 5

Charge, electric, 1, 52, 54, 69
Chu, L. J., 200
Circulation, 41
Coaxial transmission line, 130, 218, 226
Coefficient, reflection, 151, 152, 180, 184
Coil, solenoidal, 99
Commutative law, 12
Complex dielectric constant, 128
Complex quantities, 13, 125
Complex velocity, 128
Condenser, 75
  spherical, 57
Conducting material, propagation in, 125
Conduction current, 73
Conductivity, 70, 125, 133, 138, 140, 221
  earth, 184
Conductors, 53
  as boundary, 139
  electric field in, 102
  reflection from, 145
Conservation of energy, 3
Continuity, equation of, 195
Contour maps, 18
Coordinates, Cartesian, 14, 36
  cylindrical, 36, 47
  polar, 32, 36
  rectangular, 14, 36
  spherical, 36, 55
Coulomb, C. A., 61
Coulomb's law, 61
Crittenden, E. C., 9
Cross product, 11, 17
Curl, 23, 25, 26, 36, 41, inside back cover
  divergence of, 31
  of gradient, 30
Current, 69, 73, 78
  displacement, 73, 110
  surface, 142
Current density, 70
Cut-off frequency, 201, 206, 212, 214
Cylindrical coordinates, 36, 47
Cylindrical wave guides, 215

INDEX

Decibels, 130, 223, 225
Del, 25
Depth, skin, 144, 221
Depth of penetration, 144
Determinants, 17
Diamagnetism, 97
Dielectric, quasi, 128
  reflection from a, 149
Dielectric constant, 59, 138, 184
  absolute, 8
  complex, 128
  relative, 5, 8
  value of, inside front cover
Dielectric hysteresis, 129, 149
Dielectric loss, 129, 150, 224
Dielectric wave guide, 225
Dimensions, inside front cover
Direct ray, 181
Displacement current, 73, 110
Distortion, 220
Divergence, 22, 26, 36, 38, 52, inside back cover
  of curl, 31
  of gradient, 31
  of the vector potential, 162
Divergence theorem, 40
Dominant mode, 207, 218
Dorsey, N. E., inside front cover
Dot product, 10
Doublet, oscillating, 95

Earth, characteristics, 184
  reflection from, 154, 184
Effective resistance, 144
Electric charge, 1, 69
  density, 52, 54
Electric field, 1, 2, 8, 51
  in conductor, 102
Electric field strength, 8
Electrodynamic state, 107
Electromagnetic units, 78, 83
Electromagnetic waves, 110, 118
Electromotive force, 73, 80
Electron, 231, 233
Electron accelerator, 104
Electrostatic energy, 64
Electrostatic flux, 6
Electrostatic units, 5
Emde, Fritz, 215
Energy, 131, 135

Energy, conservation of, 3
  electrostatic, 64
  magnetic, 95
Equipotential surface, 45
Equivalent depth of penetration, 144
Equivalent length of antenna, 173
Everitt, W. L., 191
Experiment I, 1, 109
Experiment II, 2, 109
Experiment III, 4, 109
Experiment IV, 5, 109
Experiment V, 69, 109
Experiment VI, 77, 109
Experiment VII, 79, 109
Experiment VIII, 82, 109
Experiment IX, 83, 109
Exploring particle, 1
Exponential notation, 124, 127, 135, 147, 149

Faraday, Michael, 59, 79, 110, 117, 118
Faraday's law, 80
Ferromagnetism, 84, 96
Field, electric, 2
  electrostatic, 1, 2, 8, 51
  gravitational, 45
  induction, 167, 170
  irrotational, 46
  lamellar, 46
  magnetic, 78
  radiation, 168, 178
  scalar, 1
  solenoidal, 46
  vector, 1, 18
  vector-potential, 94
Flat-top antenna, 185
Flux, electrostatic, 6
  magnetic, 79
  linkages, 91
Flux density, electric, 6, 8, 54
  magnetic, 78, 96
Fluxmeter, 82
Force, between currents, 90
  electrostatic, 2, 51
  magnetic, 77
Franklin, W. S., 229
Frequency, cut-off, 201, 206, 212, 214
Fringing, 104, 106

Galvanometer, 79

INDEX

Gaussian units, 5, 78, 83
Gauss's theorem, 39, inside back cover
Giorgi units, 5, 8
Gradient, 19, 21, 36, inside back cover
   curl of, 30
   divergence of, 31
   potential, 45
Gravitational field, 45
Green's theorem, 40
Ground-reflected radiation, 179
Grounded antennas, 182, 189
Group velocity, 200, 203, 212, 215, 228, 236
Guide constant, 130, 211, 212, 214
Guides, cylindrical, 215
   dielectric, 225
   loss in, 220
   rectangular, 199, 205, 207
   wave, 130, 193, 198
Guillemin, E. A., 231

Hachure marks, 19
Half-wavelength antenna, 174
Heat, 46
Henry, Joseph, 79
Hertz, Heinrich, 118, 148
Hollow sphere, 61
Horn, 220
Hysteresis, dielectric, 129, 149
   magnetic, 84

Image, 182
Imaginary unit j, 125
Impedance, 125
   intrinsic, 123, 129, 151, 212, 235, 237
Index of refraction, 157, 238
Inductance, 101, 103
Induction, magnetic, 78
Induction accelerator, 104
Induction field, 167, 170
Intrinsic impedance, 123, 129, 151, 212, 235, 237
Inverse square law, 60
Ionized region, waves in, 233
Ionosphere, 231
   virtual height of, 237
Irrotational field, 46

j (imaginary unit), 125
Jahnke, Eugen, 215

Kennelly, Arthur E., 9
Kerst, D. W., 104
Kirchhoff's laws, 71, 75

Lamellar field, 46
Laplace's equation, 53, 55, 56, 160
Laplacian, inside back cover
   of scalar field, 31, 36
   of vector field, 31, 32, 36
Lenz's law, 95, 105
Light, 117, 231
   velocity of, value, inside front cover
Line, transmission, 147, 197, 231
   coaxial, 130, 218, 226
Linkages, magnetic flux, 91
Loop antenna, 189
Loss, dielectric, 129, 150
   in wave guides, 220
   power, 134, 144

Magnetic energy, 95
Magnetic field, 78
   penetration of, 140
   theories of, 96
Magnetic flux, 79
Magnetic flux density, 78, 79, 96
Magnetic flux linkages, 91
Magnetic force, 77
   between currents, 90
Magnetic induction, 78
Magnetic intensity, 85, 96
Magnetic potential difference, 100
Magnetic scalar potential, 92
Magnetic vector potential, 92
Magnetomotive force, 85, 96, 100
Maxwell, James Clerk, 61, 107, 110, 111, 117, 148
Maxwell's equations, 111, 113, 122, inside back cover
Maxwell's hypothesis, 109, 111, 117
mks units, 5, 8
Mode, dominant, 207, 218
   in wave guide, 213
   TEM, 207
   transverse electric, 207, 211
   transverse magnetic, 207, 212, 213
Moon, 238
Motion, voltage induced by, 81
Multiplication, of vectors, 10
   scalar or dot product, 10, 15

INDEX

Multiplication, triple products, 17
   vector or cross product, 11, 13, 16

Nabla, 25
Newton, Sir Isaac, 110
Newton (unit), 8, inside front cover
Notation, 4, 10, 40
   exponential, 124, 127, 135, 147, 149

Oblique reflection, 155, 180, 238
Ohm, Georg, 69
Ohm's law, 69, 74
Oscillating doublet, 95

Paddle-wheel, 23, 24, 41, 48
Parallelepiped, volume of, 17
Paramagnetism, 96
Pattern, radiation, 176, 186
Penetration, of current, 144
   of magnetic field, 140
   of waves, 128
Permeability, 83, 96, 138
   relative, 84
   value of, inside front cover
Permittivity, 8
Phase constant, 121, 128
Phase velocity, 202, 206, 212, 215, 228, 236
Plane wave, 119, 123, 168, 193
Poisson's equation, 53, 62
Polar coordinates, 32, 36
Polarization, electric, 59
   magnetic, 96
   of waves, 124
Potential, advanced, 171
   electrodynamic, 160
   electrostatic, 52
   magnetic vector, 92
   retarded, 164
   scalar, 44, 46, 92
   vector, 46, 92
Potential difference, magnetic, 100
Potential integral, 62, 93
Power, 131
   radiated, 178
Power factor, 130, 224
Poynting vector, 131, 152, 178
Practical units, 5, 8

Q (quality factor), 128
Quasi-stationary state, 94, 107, 164

Radar, 182, 220, 238
Radiated power, 178
Radiation, 107, 160, 164, 171, **172**
   ground-reflected, 179
   pattern, 176, 186
   resistance, 179, 185
Radiation field, 168, 178
Ramo, Simon, 213
Ratio, standing-wave, 153
Rationalized units, 5
Ray, direct, 181
   reflected, 181
Reactance, 75
Reactive attenuation, 215
Real part of function, 125, 135, 149
Receiving antennas, 187
Reciprocity theorem, 189
Rectangular coordinates, 14, 36
Rectangular wave guide, 199, 205, 207
Reflected ray, 181
Reflection, 138, 183, 237
   from a conductor, 145
   from a dielectric, 149
   from a semi-conductor, 153
   oblique, 155, 180, 238
Reflection coefficient, 151, 155, 158, 180, 184
Refraction, 138, 239
   index of, 157, 238
Relative dielectric constant, 5
Relative permeability, 84
Resistance, effective, 144
   radiation, 179, 185
Resistance loss, 134
Resistivity, earth, 184
   surface, 145, 221
Resonator, cavity, 219
Retarded potential, 164
Right-hand rule, 12, 88

Scalar field, 1, 18
Scalar potential, 44, 46
   magnetic, 92
Scalar product, 10, 15
Schelkunoff, S. A., 189
Sea water, 184
Semi-conductor, earth as, **184**
   reflection from a, 153
Shielding, 140, 190
Short antennas, 172

INDEX

Sign, algebraic, 87
Sinusoidal wave, 121
Skilling, H. H., 1, 147, 197
Skin effect, 141, 143, 221
Solenoidal coil, 99
Solenoidal field, 46
Specific inductive capacity, 5, 8
Spherical condenser, 57
Spherical coordinates, 36, 55
Spherical wave, 119, 168, 169, 193
Standing wave, 148, 152, 215
Standing-wave ratio, 153
Stokes' theorem, 43, inside back cover
Streamlining, 24
Surface, boundary, 138
Surface current, 142
Surface resistivity, 145, 221
Symbols, inside front cover

TEM wave, 130, 207
Temperature, 46
Terman, F. E., 130, 182, 186, 189, 191, 229
Theorem, reciprocity, 189
  uniqueness, 92
Theories of magnetic field, 96
Transmission line, 147, 197, 231
  coaxial, 130, 218, 226
Transverse electric mode, 207, 211
Transverse magnetic mode, 207, 212, 213
Transverse wave, 115, 116
Traveling wave, 112, 152, 162
Triple products, 17

Unique solution, 56
Uniqueness, theorem of, 92
Unit vectors, 13
Units, 2, 5, 8, 9, inside front cover
  cgs, 5
  electromagnetic, 78, 83
  electrostatic, 5
  Gaussian, 5, 78, 83
  Giorgi, 5, 8
  mks, 5
  practical, 5, 8

Units, rationalized, 5
  unrationalized, 5
Unrationalized units, 5

Vector analysis, 10
Vector field, 1, 18
Vector multiplication, 10
  triple products, 17
Vector potential, 46, 47
  divergence of, 162
  electrostatic, 92
  magnetic, 92
Vector product, 11, 13, 16
Vectors, unit, 13
Velocity, complex, 128
  group, 200, 203, 212, 215, 228, 236
  of light, value of, inside front cover
  phase, 202, 206, 212, 215, 228, 236
  wave, 115, 117, 123, 128
Virtual height, 237
Voltage, 58, 74, 81
Volume, 17

Water waves, 230
Wave, electromagnetic, 110, 118
  guided, 193
  in ionized region, 233
  plane, 119, 123, 168, 193
  spherical, 119, 168, 169, 193
  standing, 148, 152, 215
  traveling, 112, 152, 162
Wave equation, 113, 114, 120, 126, 143
Wave guide, 130, 193, 198
  cylindrical, 215
  dielectric, 225
  loss in, 220
  modes in, 213
  rectangular, 199, 205, 207
Wave velocity, 115, 117, 123
Wavelength, 121, 128, 212
Weber, Ernst, 9
Weber (unit), 78, 79
Whinnery, John, 213
Work, 3, 10, 58

Zigzag path of wave, 198

TABLE I

Units and Symbols

Quantity	Symbol	MKS Unit	Equivalent to:	Dimension	Page
Mechanical					
Length	s	meter	100 centimeters	L	8
Mass	m	kilogram	1000 grams	M	8
Time	t	second		T	8
Force	f	newton (joule/meter)	10^5 dynes = 102.0 grams	LMT^{-2}	8
Energy or work	W	joule (watt-second)	10^7 ergs	L^2MT^{-2}	8
Electrical					
Quantity of charge	Q,q	coulomb	3×10^9 statcoulombs *	Q	2
Electric field strength	E	volt/meter	$\frac{1}{3} \times 10^{-4}$ statvolt/centimeter *	$LMT^{-2}Q^{-1}$	8
Electric potential	V	volt	$\frac{1}{300}$ statvolt *	$L^2MT^{-2}Q^{-1}$	52
Electric flux density	D	coulomb/square meter		$L^{-2}Q$	8
Electric flux		coulomb		Q	6
Capacitance	C	farad	9×10^{11} statfarads *	$L^{-2}M^{-1}T^2Q^2$	59
Dielectric constant	ϵ	farad/meter	in free space = ϵ_0 = 8.855×10^{-12}	$L^{-3}M^{-1}T^2Q^2$	8
Relative dielectric constant	κ	numeric	ϵ/ϵ_0		8
Current	I	ampere	3×10^9 statamperes *	$T^{-1}Q$	69
Current density	ι	ampere/square meter		$L^{-2}T^{-1}Q$	70
Resistance	R	ohm	$\frac{1}{9} \times 10^{-11}$ statohm *	$L^2MT^{-1}Q^{-2}$	69
Conductivity	γ	mho/meter		$L^{-3}M^{-1}TQ^2$	70
Magnetic					
Magnetic intensity	H	ampere (turn)/meter	0.01257 oersted	$L^{-1}T^{-1}Q$	85
Magnetomotive force		ampere (turn)	1.257 gilberts	$T^{-1}Q$	85
Magnetic flux density	B	weber/square meter	10 kilogausses	$MT^{-1}Q^{-1}$	78
Magnetic flux	Φ	weber	10^8 maxwells	$L^2MT^{-1}Q^{-1}$	79
Inductance	L	henry	10^9 abhenrys	L^2MQ^{-2}	101
Permeability	μ	henry/meter	in free space = μ_0 = 1.2566×10^{-5}	LMQ^{-2}	83
Relative permeability		numeric	μ/μ_0		84
Intrinsic impedance	η	ohm	in free space = η_0 = 376.7 ohms	$L^2MT^{-1}Q^{-2}$	123

* Factors thus marked are based on $c = 3 \times 10^8$ meters/second. Values given for ϵ_0 and η_0 are based on the more accurate value (U.S. Bureau of Standards; URSI, 1972): $c = 2.9979246 \times 10^8$ meters/second.

TABLE II

Formulas and Theorems of Vector Analysis

Multiplication

Scalar Product
$$\mathbf{A} \cdot \mathbf{B} = AB \cos \angle A,B$$
$$= A_x B_x + A_y B_y + A_z B_z$$

Vector Product
$$\mathbf{A} \times \mathbf{B} = \mathbf{n}\, AB \sin \angle A,B$$
$$= \mathbf{i}(A_y B_z - A_z B_y) + \mathbf{j}(A_z B_x - A_x B_z) + \mathbf{k}(A_x B_y - A_y B_x)$$
$$= \begin{vmatrix} \mathbf{i} & \mathbf{j} & \mathbf{k} \\ A_x & A_y & A_z \\ B_x & B_y & B_z \end{vmatrix}$$

Differentiation

Rectangular Coordinates (Mutually perpendicular unit vectors $\mathbf{i}$, $\mathbf{j}$, $\mathbf{k}$.)

Gradient
$$\nabla P = \mathbf{i}\frac{\partial P}{\partial x} + \mathbf{j}\frac{\partial P}{\partial y} + \mathbf{k}\frac{\partial P}{\partial z}$$

Divergence
$$\nabla \cdot \mathbf{A} = \frac{\partial A_x}{\partial x} + \frac{\partial A_y}{\partial y} + \frac{\partial A_z}{\partial z}$$

Curl
$$\nabla \times \mathbf{A} = \begin{vmatrix} \mathbf{i} & \mathbf{j} & \mathbf{k} \\ \frac{\partial}{\partial x} & \frac{\partial}{\partial y} & \frac{\partial}{\partial z} \\ A_x & A_y & A_z \end{vmatrix} = \mathbf{i}\left(\frac{\partial A_z}{\partial y} - \frac{\partial A_y}{\partial z}\right) + \mathbf{j}\left(\frac{\partial A_x}{\partial z} - \frac{\partial A_z}{\partial x}\right)$$
$$+ \mathbf{k}\left(\frac{\partial A_y}{\partial x} - \frac{\partial A_x}{\partial y}\right)$$

Laplacian
$$\nabla^2 F = \nabla \cdot \nabla F = \frac{\partial^2 F}{\partial x^2} + \frac{\partial^2 F}{\partial y^2} + \frac{\partial^2 F}{\partial z^2}$$
$$\nabla^2 \mathbf{A} = \mathbf{i}\,\nabla^2 A_x + \mathbf{j}\,\nabla^2 A_y + \mathbf{k}\,\nabla^2 A_z$$

Cylindrical Coordinates (Mutually perpendicular unit vectors $\mathbf{1}_r$, $\mathbf{1}_\theta$, $\mathbf{k}$.)
$$x = r \cos \theta \qquad y = r \sin \theta \qquad z = z$$

Gradient
$$\nabla P = \mathbf{1}_r \frac{\partial P}{\partial r} + \mathbf{1}_\theta \frac{1}{r}\frac{\partial P}{\partial \theta} + \mathbf{k}\frac{\partial P}{\partial z}$$

Divergence
$$\nabla \cdot \mathbf{A} = \frac{\partial A_r}{\partial r} + \frac{A_r}{r} + \frac{1}{r}\frac{\partial A_\theta}{\partial \theta} + \frac{\partial A_z}{\partial z}$$

Curl
$$\nabla \times \mathbf{A} = \begin{vmatrix} \frac{\mathbf{1}_r}{r} & \mathbf{1}_\theta & \frac{\mathbf{k}}{r} \\ \frac{\partial}{\partial r} & \frac{\partial}{\partial \theta} & \frac{\partial}{\partial z} \\ A_r & rA_\theta & A_z \end{vmatrix}$$

(In 2 variables: $\nabla \times \mathbf{A} = \mathbf{k}\left(\dfrac{\partial A_\theta}{\partial r} - \dfrac{1}{r}\dfrac{\partial A_r}{\partial \theta} + \dfrac{A_\theta}{r}\right)$)

Laplacian
$$\nabla^2 F = \frac{\partial^2 F}{\partial r^2} + \frac{1}{r}\frac{\partial F}{\partial r} + \frac{1}{r^2}\frac{\partial^2 F}{\partial \theta^2} + \frac{\partial^2 F}{\partial z^2}$$
$$\nabla^2 \mathbf{A} = \mathbf{1}_r\left(\nabla^2 A_r - \frac{2}{r^2}\frac{\partial A_\theta}{\partial \theta} - \frac{A_r}{r^2}\right) + \mathbf{1}_\theta\left(\nabla^2 A_\theta + \frac{2}{r^2}\frac{\partial A_r}{\partial \theta} - \frac{A_\theta}{r^2}\right)$$
$$+ \mathbf{k}\,(\nabla^2 A_z)$$

Spherical Coordinates (Mutually perpendicular unit vectors $\mathbf{1}_r$, $\mathbf{1}_\theta$, $\mathbf{1}_\phi$)

$$x = r \cos\phi \sin\theta \qquad y = r \sin\phi \sin\theta \qquad z = r \cos\theta$$

Gradient $\qquad \nabla P = \mathbf{1}_r \dfrac{\partial P}{\partial r} + \mathbf{1}_\phi \dfrac{1}{r \sin\theta} \dfrac{\partial P}{\partial \phi} + \mathbf{1}_\theta \dfrac{1}{r} \dfrac{\partial P}{\partial \theta}$

Divergence $\nabla \cdot \mathbf{A} = \dfrac{1}{r^2} \dfrac{\partial}{\partial r}(r^2 A_r) + \dfrac{1}{r \sin\theta} \dfrac{\partial A_\phi}{\partial \phi} + \dfrac{1}{r \sin\theta} \dfrac{\partial}{\partial \theta}(A_\theta \sin\theta)$

Curl $\qquad \nabla \times \mathbf{A} = \begin{vmatrix} \dfrac{\mathbf{1}_r}{r^2 \sin\theta} & \dfrac{\mathbf{1}_\theta}{r \sin\theta} & \dfrac{\mathbf{1}_\phi}{r} \\ \dfrac{\partial}{\partial r} & \dfrac{\partial}{\partial \theta} & \dfrac{\partial}{\partial \phi} \\ A_r & r A_\theta & r \sin\theta\, A_\phi \end{vmatrix}$

Laplacian $\qquad \nabla^2 F = \dfrac{\partial^2 F}{\partial r^2} + \dfrac{1}{r^2} \dfrac{\partial^2 F}{\partial \theta^2} + \dfrac{1}{r^2 \sin^2\theta} \dfrac{\partial^2 F}{\partial \phi^2} + \dfrac{2}{r} \dfrac{\partial F}{\partial r} + \dfrac{\cot\theta}{r^2} \dfrac{\partial F}{\partial \theta}$

Theorems

Gauss's $\qquad\qquad \oint \mathbf{E} \cdot d\mathbf{a} = \int \nabla \cdot \mathbf{E}\, dv$

Stokes' $\qquad\qquad \oint \mathbf{E} \cdot d\mathbf{s} = \int (\nabla \times \mathbf{E}) \cdot d\mathbf{a}$

Lamellar Field (Irrotational Field) and *Scalar Potential*

$$\nabla \times (\nabla P) \equiv 0$$

and conversely if $\qquad \nabla \times \mathbf{B} = 0$
it is possible to let $\qquad \mathbf{B} = -\nabla P$

Solenoidal Field (Sourceless Field) and *Vector Potential*

$$\nabla \cdot (\nabla \times \mathbf{A}) \equiv 0$$

and conversely if $\qquad \nabla \cdot \mathbf{B} = 0$
it is possible to let $\qquad \mathbf{B} = \nabla \times \mathbf{A}$

TABLE III

ELECTROMAGNETIC EQUATIONS

Equation Number		
240	$\nabla \times \mathbf{H} = \iota + \dfrac{\partial \mathbf{D}}{\partial t}$	$\mathbf{D} = \epsilon \mathbf{E}$
193	$\nabla \times \mathbf{E} = -\dfrac{\partial \mathbf{B}}{\partial t}$	$\mathbf{B} = \mu \mathbf{H}$
196	$\nabla \cdot \mathbf{B} = 0$	$\iota = \gamma \mathbf{E}$
115	$\nabla \cdot \mathbf{D} = \rho$	

WAVE RELATIONS

$$\beta \lambda = 2\pi \qquad\qquad \lambda = \dfrac{v}{f} = \dfrac{2\pi v}{\omega}$$

$$\beta = \dfrac{2\pi}{\lambda} = \dfrac{\omega}{v} \qquad\qquad v = \dfrac{\omega}{\beta}$$

$$\lambda = \dfrac{2\pi}{\beta} \qquad\qquad v_g = \dfrac{d\omega}{d\beta}$$

Note: v is phase velocity and may be written v_ϕ; v_g is group velocity.

St. John's College
Annapolis, Maryland
January 1980